"*Origins* presents a perspective on origins of the universe, the
earth, life on earth, and of huma[...]
to science. In an easy-to-underst[...]
provide a valuable resource for s[...]
interested in origins."

D0562995

[...]ac

Executive Direc[...]on

"The Haarsmas provide a comprehensive perspective on science
and its challenges viewed through a lens of firm belief in God,
which illuminates the issues and offers straight ways forward.
Christians will be encouraged by their faithful and convincing
approach."

Daniel M. Harrell
Senior Minister, Colonial Church, Edina, Minnesota;
author of Nat*ure's Witness: How Evolution Can Inspire Faith*

"Like Jacob's experience, this book helps us wrestle with God until
we are blessed. It engages Scripture and science in a way that does
not dismiss or distort either, and so it builds a more comprehensive
appreciation of the mystery and creation of God. Employing
elements in the Reformed tradition of reasoning and a survey of
contemporary Christian beliefs, it helped me understand the views
of others that differ from mine, while helping me understand why I
believe as I do."

Dr. Joel C. Hunter, Senior Pastor
Northland Church

REVISED EDITION

DEBORAH B. HAARSMA & LOREN D. HAARSMA

ORIGINS

CHRISTIAN
PERSPECTIVES
ON CREATION,
EVOLUTION,
AND
INTELLIGENT
DESIGN

FAITH
ALIVE®
Christian Resources

Grand Rapids, Michigan

Origins: Christian Perspectives on Creation, Evolution, and Intelligent Design, © 2011 by Faith Alive Christian Resources, Grand Rapids, Michigan.

Printed in the United States of America.

We welcome your comments. Call us at 1-800-333-8300, or e-mail us at editors@faithaliveresources.org.

Library of Congress Cataloging-in-Publication Data
Haarsma, Deborah B.
    Origins : Christian perspectives on creation, evolution, and intelligent design / Deborah B. Haarsma & Loren D. Haarsma.
    — [2nd ed.].
        p. cm.
    Includes bibliographical references and index.
    ISBN 978-1-59255-573-4
    1. Creation. 2. Cosmogony. 3. Beginning. 4. Biblical cosmology. 5. Evolution (Biology)—Religious aspects—Christianity. 6. Christian Reformed Church—Doctrines. I. Haarsma, Loren D. II. Title.
    BS651.H33 2011
    231.7'652—dc22

                                                    2011010602

10 9 8 7 6 5 4 3 2

CONTENTS

ACKNOWLEDGMENTS

We wish to thank our family, friends, pastors, and teachers who have shown us over the years, by word and deed, what it means to love God. Thank you for encouraging us to study science and to use our gifts in Christ's service.

We are grateful to our friends and family members who patiently read the much longer early draft of this work and encouraged us in this project: David Becker, Eileen Becker, Wilbert Becker, Gayle Deters, Beth Haarsma, Henry Haarsma, Dan Hawkins, Sean Ivory, Art Mulder, Larry Osterbaan, Jack Snoeyink, and Ben VanHof.

Thanks also to the scientists, theologians, and other experts who read early drafts and shared their knowledge and insights: Lyle Bierma, Amy Black, Joy Bonnema, John Cooper, Ronald Feenstra, Stan Haan, Thea Leunk, Steven Matheson, Clarence Menninga, Glenn Morton, Steve Moshier, Craig Rusbult, Ralph Stearley, Daniel Treier, Jeremy VanAntwerp, and Davis Young.

Finally we are grateful to the editors and staff at Faith Alive Christian Resources for inviting us to write this book, and for their unflagging commitment to aid Christians who want to learn about creation, design, and evolution. We are honored to be a part of this project.

—Deborah B. Haarsma and Loren D. Haarsma

PREFACE TO THE REVISED EDITION

The first edition of this book, *Origins: A Reformed Look at Creation, Design, and Evolution* (2007), was commissioned for a target audience of Christians familiar with Reformed theology and tradition. This new edition is intended for a broader audience. The content is substantially unchanged. In various places we have modified the text, corrected a few mistakes in the original edition, and removed most "insider" references to the Reformed tradition. We moved the topic of fine-tuning from chapter 10 to chapter 7, added answers to some common questions to chapter 13, and added a new chapter on worship. We have also updated the lists of Additional Resources to include recent books and articles.

Many Christians consider the topic of origins "dangerous waters" —if you sail in the wrong direction, you could run your intellect aground or, worse yet, shipwreck your faith. Our goal is to help you navigate those waters. In some parts, especially the early chapters, we steer through broad channels where Christians of many traditions generally have consensus. In other parts, especially the later chapters, we venture into rocky areas where Christians disagree with each other, sometimes sharply. Rather than trying to steer you in one set direction, we chart out several paths that Christians take while pointing out some of the hazards along the way.

—Deborah B. Haarsma and Loren D. Haarsma
January 2011

INTRODUCTION

Creation. *Evolution. Design.* These words mean different things to different people, and they tend to provoke arguments. The issues they represent run close to some of our deepest questions:

▶ Does God exist?
▶ How does God relate to this universe?
▶ How did we get here?
▶ Who are we?

Among Christians, opinions on creation, evolution, and design vary widely. We've seen this variety as we've traveled around the country to speak to churches, to pastors, and to Christian students at colleges and universities. The Christians we meet share a common belief about *who* created the universe. But they believe very different things about *how* God created the universe.

Some Christians we know are eager to see scientific proof to support their belief that God created everything. Others believe that there can't be any scientific proof, because if we have proof then we don't need faith. Some feel that the Bible is only about spiritual issues and is irrelevant to science. Others claim that the Bible tells us scientific information, including the age of the earth and the beginning of life. Some Christians believe that the earth was created just a few thousand years ago, and they are suspicious when they hear about the Big Bang theory. They dislike the very word *evolution* and wonder how any Christian could believe that we evolved from apes. Some Christian students have even been warned that they will be in danger of losing their faith if they listen to scientific evidence for evolution. But other Christians feel that believing in the Bible is perfectly compatible with believing that the earth is old and that God used evolution to create life; they wonder why Christians are still wasting their time arguing about these issues. Yet other Christians are somewhere in between. They have been taught that it's acceptable to believe in an old earth, but they're not so sure about biological evolution, and they wonder whether gaps in evolution show that an intelligent designer must have brought about life on the earth.

Religion and Science in Conflict?

Some people see these arguments about creation, design, and evolution as a conflict with just two sides:

▶ Atheists on one side who try to use science to disprove religion.

▶ Religious believers on the other side who reject the work of scientists.

If there are only two sides to choose from, then believing what science says means rejecting God, and believing in God means rejecting science. Which do you choose: science or religion?

Of course it's not that simple. The issues around creation, evolution, and design are more complex—and more interesting.

> Throughout this book we will use the word *science* to refer collectively to the natural sciences such as physics, astronomy, chemistry, geology, biology, and so on. Social sciences such as sociology, political science, and economics are not the focus of this book.

Debate the Weather?

To illustrate why the debate about origins isn't simply a matter of science versus religion, imagine living in a culture where there is a similar debate about the weather. The Bible clearly teaches that God governs the weather. Many Bible passages proclaim that God causes rain and drought (see Deut. 11:14-17; 1 Kings 8:35-36; Job 5:10; 37:6; Jer. 14:22). Writers of Deuteronomy, the Psalms, and Jeremiah refer specifically to storehouses of rain and snow (see Deut. 28:12, 24; Ps. 135:7; Jer. 10:13).

What causes the rain? Most of us were taught that water evaporates from the ground level, rises to where the air is cooler, and condenses into water droplets that form clouds. We learned how cold fronts and warm fronts and low pressure systems bring rain. When we watch meteorologists on television, we hear that scientists now use sophisticated computer models to help them understand and predict the weather a few days in advance. Their ability to understand meteorology is especially important for farmers, airline pilots, military personnel, and coastal residents. Every year scientists develop increasingly accurate computer models of the weather.

Now imagine that debates arise about what should be taught in schools about the weather. Imagine that prominent scientists write popular books about meteorology that state, "From our scientific understanding of the causes of wind and rain, it is clear that no divine being controls the weather." Imagine that a professional organization of science teachers writes a set of guidelines that state, "Students must learn that all weather phenomena follow from natural causes; weather is unguided and no divine action is involved." Meanwhile, other people insist that these scientific explanations of rain and wind must be wrong because the Bible clearly teaches that God governs the weather. These people write books and give public speeches saying, "Atheists have invented their godless theories about evaporation and condensation. But we can prove that their so-called scientific theories are false and that the Bible is true." They go to churches and teach, "If you believe what these scientists are saying about the causes of wind and rain, then you've abandoned belief in the Bible." They petition school boards and courts to require that science class-rooms also teach their "storehouses" theory of the weather as an alternate explanation to evaporation and condensation.

If you lived in a world with that sort of debate going on, would you be content to see it simply as a conflict between science and religion? Would you be willing to agree wholly with one side or the other?

More Than Two Options

Fortunately, we don't have such debates about what causes the weather. The majority of Christians say that when it comes to the weather, both science and the Bible are correct. God governs the weather, usually through the scientifically understandable processes of evaporation and condensation. And the majority of atheists today would also agree that having a scientific explanation for the weather, by itself, neither proves nor disproves the existence of God. So there are no court battles about what science classrooms should teach about the weather.

Debates about creation, evolution, and design have some similarities to the above example, but in many ways they are more difficult. The questions about how to interpret Scripture are more challenging, and these debates raise more theological issues. A good place to start in making sense of these debates is to remember that more than two options exist.

About This Book

The purpose of this book is to lay out a wider variety of options and to examine what both the Bible and the natural world can teach us about these options. We will explore in depth the issues of origins and consider areas where Christians generally agree with each other *and* areas where Christians disagree. Our goal is not to convince you that one particular opinion must be correct, but neither will we merely list a wide variety of opinions without doing any analysis. We will

▶ summarize what we believe God's *Word* teaches about origins when it is studied using sound principles of interpretation.

▶ summarize what we believe God's *world* can reliably reveal about origins when it is studied using sound scientific methods.

▶ distinguish between scientific theories that are well established and have a great deal of data supporting them and scientific theories that are more tentative and speculative.

▶ look at the range of opinions that Christians hold about origins and discuss some of the pros and cons of each in light of what we can learn from God's Word and God's world.

This book often refers to Christians as *we/us* because it is written primarily for people who believe that the Bible has been inspired by God and who choose to live according to its teaching. The primary audience is people who believe that God created the universe and that a scientific study of God's world and a careful study of God's Word both have something to teach us about origins. We will not try to *prove* these beliefs, since we expect that most readers will share these assumptions. If you don't share these assumptions, this book at least will give you some insight into people who do.

The first chapter of this book looks at the relationship between the study of God's Word and God's world. Rather than placing theology over science or science over theology, we will explore the sovereignty of God over both.

Chapter 2 argues that the practice of science is consistent with a Christian worldview and that Christians can work with scientists of other worldviews without compromising their Christian beliefs. We then consider God's governance in four areas: explainable natural events, unexplainable natural events, supernatural miracles, and random events.

Chapter 3 examines three methods (experimental, observational, and historical) used to gain scientific knowledge about God's world. Historical science has proven to be a reliable method for learning about natural history. The chapter closes with a discussion of the limits of science.

Chapter 4 describes science and theology as the human interpretations of God's two revelations, nature and Scripture. We consider factors that influence these human interpretations and discuss when these interpretations can be considered reliable. As a case study, we consider Galileo's conflict with the church regarding the motion of the earth through space. Galileo's story holds several important lessons for understanding today's conflicts.

Chapters 5 and 6 examine what nature and Scripture are telling us about the age of the earth. Chapter 5 describes the geological evidence for age, as well as four *concordist* interpretations of Genesis 1. Chapter 6 describes five *non-concordist* interpretations of Genesis 1 and then discusses all of these interpretations in light of the principles of biblical interpretation set forth in chapter 4.

Chapter 7 looks at what astronomers are learning about the history of the universe. We'll describe the evidence that the universe is incredibly vast, with a long, dynamic history. It is old, but the evidence for the Big Bang shows that it is not *infinitely* old; it had a beginning in time. Many scientists today are saying that the fundamental laws of nature have been "fine-tuned" for life to exist and develop around stars.

Chapter 8 sorts out the various meanings of the word *evolution* and takes a closer look at the atheistic worldview of *evolutionism*. It defines the *progressive creation* and *evolutionary creation* positions and discusses some theological issues they raise.

Focusing on plant and animal evolution, chapter 9 describes the scientific evidence from fossils, anatomy, geography, and genetics in support of common ancestry and the theory of evolution. It concludes with an analysis of three Christian positions on origins in light of this evidence.

Chapter 10 discusses how the term *Intelligent Design* is being used in today's debates over origins. It compares fine-tuning arguments to Intelligent Design arguments and considers in detail the question of how biological life became so complex.

Chapter 11 summarizes the scientific evidence regarding human origins and discusses relevant theological topics, such as original sin and what it means that humans are made in the image of God. Chapter 12 considers five scenarios for Adam and Eve and when they lived, examining the pros and cons of each scenario in light of the scientific and theological issues raised in chapter 11. The Appendix lists over a dozen positions on creation, design, and evolution origins that are consistent with belief in God.

Chapter 13 summarizes many of the questions Christians ask about origins and discusses a few questions in particular: proofs of God in nature, human significance, and worship in the context of origins. It concludes with some advice for how Christians can live and work together despite their differing views on creation, evolution, and design.

In Chapter 14 and throughout this book you'll find places where we pause to praise the Creator for the wonder and beauty of the world he has made. Chapter 14 discusses worship and its importance in the midst of these debates.

STUDY TIPS

Designed for individual or small group study, each chapter of this book includes

▶ an in-depth look at each topic.
▶ questions for reflection or discussion.
▶ a list of suggested resources for further reading and study.

If you plan to use this book in a study group, the chapters can be combined into six, four, or even three sessions. For six sessions, we recommend grouping as follows: chapters 1-2, 3-4, 5-6, 7, 8-10, and 11-12. For four sessions, you can group chapters 1-4, 5-7, 8-10, and 11-12. For three sessions, group chapters 1-4, 5-7, and 8-12. The Introduction, chapter 13, chapter 14, and the Appendix can be used as supplements.

If you are new to leading a discussion group, here are some suggestions:

▶ Encourage everyone to read before coming so that all will be informed.

▶ Start the meeting by asking people to share one or two things they found particularly interesting in the reading. Depending on your group, that may be all you need to get the discussion started.

▶ Allow everyone to share their views, but if one or two people begin to dominate the discussion you can direct conversation to another aspect of the reading or invite someone who has been quiet to share his or her thoughts.

▶ If conversation lags or bogs down in one topic, use the questions at the end of the chapter to get the discussion moving (look them over ahead of time to pick the questions you think will work well with your group).

▶ When necessary, remind people not to say negative things about other Christians (don't suggest that others are stupid or lacking in faith); critique ideas, not people.

▶ End with a song, Scripture reading, or prayer to praise God for his handiwork and remind everyone of our unity in Christ.

If you or members of your study group wish to pursue the topic of origins even further, you'll find a collection of over forty related articles on our website (www.faithaliveresources.org/origins). Throughout the chapters in this book, you'll find sidebars directing you to specific articles on our website as they relate to the topic at hand.

Don't worry if it's been a while since you last sat in a science classroom—we expect that most readers are not scientists or theologians. Although the reading will get a bit technical in places, we'll try to avoid using too much jargon. It is our hope that your study of this book will give you a greater understanding of the complex topic of origins and a deeper sense of awe for our great God.

About the Authors

I grew up loving science and math, as well as music and reading. My teachers at school and at church encouraged my academic interests. In the evangelical church we attended I was taught a young earth interpretation of Genesis 1, but this view was not emphasized as an essential article of faith. In my public high school evolution was taught in biology class, but, fortunately, not in an antireligious way.

I attended Bethel College in St. Paul, Minnesota, majoring in physics so that I could combine my love of science with my love of math. It was amazing to discover how complex calculations could accurately match what happened in real-world experiments. I learned that science is founded on a Christian worldview in that scientific investigation relies on the regularity of natural laws, which in turn rely on God's faithful governance. One day I heard a chapel speaker talk about the need for more Christians at universities to bring the gospel to academics and a Christian worldview to academic culture. I felt God's call that day to be a Christian voice among scientists and to be a scientific voice within the church. I still follow this call, eager to bridge the gap and show both scientists and Christians that the two are not in opposition.

In graduate school my general interest in physics focused into a particular interest in astrophysics. I earned a PhD at the Massachusetts Institute of Technology, doing research on galaxies and the expansion of the universe. As my research project turned to the question of the age of the universe, I realized that I needed to learn more about how Christians understood issues of age and origins. I read several authors and discussed the issue with fellow Christian graduate students in my InterVarsity chapter. I came to understand that a high view of Scripture can agree with a deep respect for science and what it tells us about God's world.

At a scientific conference I attended as a graduate student, I heard an astronomer speak about the importance of sharing the latest astronomical results with the general public. Scientists should not only report their findings to each other but to all the children and adults who would like to know more about the universe—particularly if those scientists are funded by taxpayer dollars. I realized that, as good as that reason was, I had a deeper and better reason for telling people about astronomy. It is the very handiwork of God. As I learn about physics in the universe, it is not enough for my own heart to rise in praise. I need to tell it to others so that they too can praise God for his glory declared in the heavens. In my teaching at Calvin College and in this book, God has given me more opportunities to tell of his handiwork.

—Deborah B. Haarsma

I have always loved learning science. I remember reading books on astronomy when I was still in the early grades of elementary school. And I have always loved learning theology. Since I grew up in the Christian Reformed Church and attended Christian schools from kindergarten through college, I had many opportunities to learn theology. My earliest memory of bringing together science and theology is from a Sunday school class in middle school. The pastor talked about Scripture's teaching that God keeps all of the planets in their proper place. He also pointed out that science can explain that planets stay in their orbits because of gravity. Then he assured us that these two views are not in conflict: God created the law of gravity and uses it to keep the planets in orbit. That incident—and the Reformed theology behind it—has guided my thinking and calmed my worries over the years whenever someone claims that science and religion are in conflict. Another great help to me was a Bible class during my senior year of high school. I don't recall discussing origins or any other scientific issue in that class, but we did learn a great deal about interpreting, understanding, and applying Scripture.

Growing up, I was usually taught a young earth interpretation of Genesis; however, I don't recall any teachers or pastors insisting that this was the only right interpretation. In high school and early college classes I began to learn the scientific evidence related to the age and history of the earth. I was aware that some Christians mistrusted this science, but I knew from my studies that scientists were usually very smart and very careful. Reading what a variety of theologians had written about interpreting Genesis deepened my understanding. Their writings helped me to reconcile the study of God's world with the study of God's Word.

Since my graduation from Calvin College, science has been my career. I attended graduate school at Harvard University and earned my PhD in physics. Then for five years I did full-time scientific research in the exciting field of neuroscience. Neuroscience combines a biologist's study of how cells and living organisms work with a physicist's study of how atoms move and electrical signals travel.

In 1999 I returned to Calvin College to teach. Throughout my tenure there my love for both science and theology has grown. Whenever someone claims that there is a conflict between science and Christian theology on some particular issue, I feel called, compelled, and eager to explore that issue. Again and again over the years I have found, by carefully studying the science, philosophy, and theology surrounding the issue, that no contradiction really exists. Not only can science and theology be reconciled, but each can serve to increase our understanding of the other.

—Loren D. Haarsma

CHAPTER 1

GOD'S WORD AND GOD'S WORLD

"What is God like?" Questions like this are essentially religious and not scientific. "What is the mass of a carbon atom?" Questions like this are essentially scientific and not religious. It might seem that the simplest way to avoid conflict between religion and science is to keep them completely separate from each other. A number of atheists and agnostics have written in favor of this idea. They think that people should look to their religion or personal philosophy to answer questions about moral meaning and personal value—questions that science *cannot* answer, and they ask in return that religion defer to science on questions about the natural world—questions that science *can* answer. It's a tempting idea.

But it is not always possible to keep science and religion in separate compartments. Some discoveries of science inevitably raise religious questions. Here are just a few examples:

▶ When astronomers study the vastness of the universe, we can't help but wonder about the significance of tiny humans in such a large universe.

▶ When biologists study organisms that cause disease, we ponder the causes of suffering.

▶ When ecologists study the great variety of plants and animals in this world, we reconsider our responsibility to take care of this world.

▶ When physicists discover powerful new energy sources, we debate whether they should be used for war.

Scientific discoveries like these certainly affect our beliefs and decisions about moral and religious questions.

Most people, Christians or atheists, are not content to hold competing and contradicting beliefs in separate compartments. We don't want to have one set of beliefs when we study the natural world, another set when we decide how to vote, a third set when we decide how to spend our money, and a fourth set for church. We want all the parts of our lives to flow from a unified, consistent set of beliefs.

Most importantly, Christians cannot simply separate science from religion because the Bible proclaims that God is sovereign over every part of life. The God who created the planets and the stars is also the God who inspired the Bible and who is personally revealed in human history. The God who made the sky and the ocean is also the God who commands us to act out of love rather than selfishness. The God who made the plants and animals is also the God who redeems us after we disobey his commands. The God who gave us the ability to study the world scientifically is also the God who guides us with the Holy Spirit as we seek to understand his written revelation. We cannot separate our study of God's Word from our study of God's world because both come from and point us toward the same God.

CHRISTIANS AGREE AND DISAGREE

Christians in Agreement

When Christians discuss creation, evolution, and design, it is easy to focus immediately on areas of controversy and disagreement. We think it is important to start by pointing out certain areas in which nearly all Christians agree and which we strongly affirm in this book. Christians generally agree about the fundamentals of God, God's Word, and God's world in the five areas described below.

▶ **God created, sustains, and governs this universe.**

This truth is confirmed in the first line of the Apostles' Creed, one of the ecumenical creeds of the church that many Christians recite every week: "I believe in God, the Father almighty, creator of heaven and earth." Christians believe that God created all things from nothing, bringing them into being through his Word, his Son (John 1:1-3). God continually sustains the whole universe, governing all creatures according to his providential care.

▶ **The God who created this world also reveals himself to humanity.**

God has revealed himself at various times and in multiple ways throughout history, including the written Scriptures and the Incarnation. As it says in the first verses of the book of Hebrews,

> In the past God spoke to our ancestors through the prophets at many times and in various ways, but in these last days he has spoken to us by his Son, whom he appointed heir of all things, and through whom also he made the universe. The Son is the radiance of God's glory and the exact representation of his being, sustaining all things by his powerful word. After he had provided purification for sins, he sat down at the right hand of the Majesty in heaven.
>
> —Hebrews 1:1-3, NIV.

▶ **The God who created this world is also our Redeemer.**

We belong to God because he created us, but when humanity turned from God he bought us back. He redeemed us through the incarnation, life, suffering, death, and resurrection of Jesus Christ.

▶ **The Bible is authoritative and sufficient for salvation.**
God inspired its human authors and ensured that the Bible truthfully teaches what he intends. The Holy Spirit testifies in our hearts that the Bible's message is from God, not merely human writing. Christians accept the sufficiency of the Bible for establishing our core beliefs and practices; all that we need to know for salvation is taught there. God certainly can use various means—including the natural world—to teach us new things. But these new things should be compatible with, not contradictory to, what God teaches in Scripture.

▶ **God is sovereign over all realms of human endeavor and has given human beings special abilities and responsibilities. Theologian Cornelius Plantinga puts it this way:**
God's creation extends beyond the biophysical sphere to include the vast array of cultural possibilities that God folded into human nature. . . . God's good creation includes not only earth and its creatures, but also an array of cultural gifts, such as marriage, family, art, language, commerce, and (even in an ideal world) government. The fall into sin has corrupted these gifts but hasn't unlicensed them. The same goes for the cultural initiatives we discover in Genesis 4, that is, urban development, tent-making, musicianship, and metal-working. All of these unfold the built-in potential of God's creation. All reflect the ingenuity of God's human creatures—itself a superb example of likeness to God.
—Cornelius Plantinga, *Engaging God's World,* 2002.

Applying this idea to the natural sciences, we conclude that God has graciously given humans the ability and responsibility to study the natural world systematically. As with all human endeavors, we do it imperfectly. We must seek to do it as God's imagebearers, in gratitude for God's gifts.

Christians in Disagreement

Christians have always agreed about *who* created everything, but in the last few decades they have often disagreed about *how* God created everything. These disagreements are over two basic questions:

▶ As we study God's Word, what is the best way to understand passages that talk about God's acts of creation?

▶ As we study God's world, what can we reliably conclude that it tells us about its history?

Some Christians describe themselves as *young-earth creationists*. They believe that the best interpretation of the book of Genesis is that the earth is only a few thousand years old and was shaped by a global flood. Young-earth creationists hold a range of views about how to interpret Scripture, the extent to which scientific data indicates a young universe, and the extent to which it indicates at least an appearance of long history.

Other Christians describe themselves as *old-earth creationists*. Some believe that in the best interpretation of Genesis 1, the events on each day actually describe several long epochs of scientific history. Others believe that the best interpretation of the book of Genesis does not imply anything about the age of the earth one way or the other and that drawing conclusions about the age of the earth from Scripture is reading into it something it was never intended to teach.

Some old-earth creationists describe themselves as *evolutionary creationists*. They believe that the best understanding of the scientific data—in conjunction with the best interpretation of Scripture—implies that God governed and used evolutionary processes in the unfolding of creation. Other old-earth creationists describe themselves as *progressive creationists*. They believe that science and Scripture both indicate that God used not only natural processes but also some miracles along the way, particularly in the history of life. Arguments for *Intelligent Design* are usually, though not always, used to support versions of progressive creation.

One purpose of this book is to examine this range of young-earth and old-earth beliefs and the various ways in which Christians think about creation, design, and evolution. We will define these terms and explore that range of beliefs in more detail in chapters 5, 6, and 8, as well as in the Appendix.

Because of these disagreements among Christians, a number of churches, denominations, and organizations have affirmed that Christians can and do hold a variety of views on origins—each motivated by a sincere desire to be faithful to God and to Scripture—and that a range of views falls within the bounds of Christian belief. (See the notes at the end of this chapter for more examples of positions taken by various denominations and Christian organizations.)

LISTENING TO BOTH THE WORD AND THE WORLD

Whenever Christians discuss origins, sooner or later someone asks the question "Are you going to let science tell you how to read the Bible, or are you going to let the Bible tell you how to do science?" Instead of asking whether theology should dictate science, or science theology, we suggest asking the following questions:

▶ Is it ever appropriate for Christians to allow what we learn from the study of creation to affect how we interpret Scripture?
▶ Is it ever appropriate for Christians to allow what we learn from the study of Scripture to affect how we interpret creation?

We think the answer to both questions is yes but that we can do each in good ways and bad ways.

Scripture's Influence on Interpretation of Nature
An inappropriate use of the Bible is to read *into* Scripture things it was never intended to teach. While Scripture is authoritative and

sufficient to teach us everything we need to know to be saved, it is not intended to be a reference book covering all human knowledge. The Bible has things to say about how we should farm, but it is not a farming manual. The Bible has things to say about how we use our money, but it is not an economics text. The Bible has things to say about the natural world, but it is not a science text. We wouldn't expect a scientist to use the Bible to figure out how an electronic circuit responds to signals, just as we wouldn't expect a plumber to use the Bible to figure out what size pipe to use in a house under construction.

But Scripture can influence the interpretation of the natural world in appropriate ways. For example, the Bible teaches us that God can interact with the natural world. Scientists are trained to study the natural world and try to explain as much as they can in terms of natural processes. If a scientist spent his whole life only looking to science for answers, he might be tempted to conclude that every event must *always* have a purely natural explanation. Scripture offers a correction here. Scripture teaches that God can and does sometimes work outside of normal, natural processes. God can do miracles. Thus, a Christian scientist must be open to the possibility that some events might not have scientific explanations.

The Bible teaches us important lessons about the natural world that science alone cannot discover. Scientists have discovered that human bodies are made of the same types of atoms as plants and rocks and dirt; nevertheless, the Bible teaches that humans have a special place in creation. Scientists can explain how spring, summer, autumn, and winter are caused by the angle of earth's rotation relative to its orbit around the sun; the Bible teaches that the seasons follow each other because of God's faithful, continuing governance.

Science's Influence on Interpretation of Scripture

Is it appropriate to allow science to influence the interpretation of Scripture? Some influences would *not* be appropriate. We shouldn't use science as an excuse to ignore parts of the Bible that seem to conflict with science. Scientific reasoning alone should not drive how we interpret the Bible. Science should not push us to interpret one part of the Bible in ways that conflict with the clear teachings of other parts of the Bible.

But some influences can be appropriate. Things that we learn from studying God's world can help us to better understand some parts of God's Word. Think of this in a broader context. All of life's experiences have the potential to alter and improve our understanding of Scripture. For example, if we mistakenly interpreted Scripture to justify a "health and wealth" gospel (the premise that God blesses us with riches and good health so long as our faith is strong and we don't sin too much), then God might use life's experiences of suffering and loss, as well as the wisdom of other Christians, to correct our mistaken interpretation. Similarly, God can use scientific encounters with his world to help us better understand parts of Scripture.

Science can be helpful in cases where the Bible itself is ambiguous and open to multiple interpretations. Think about ordinary events like sunrise and sunset, regular patterns of rainfall, and seeds growing in a field of tall corn. Are these events that happen all the time, brought about miraculously by God? Or does God bring about these events using natural laws and ordinary processes, without the need for his miraculous intervention each time? Many Bible passages speak about God establishing night and day, about God bringing rainy and dry seasons, about God blessing the land with good harvest. It would be easy to conclude from these passages that God uses special, miraculous action to cause these everyday events.

But it is also easy to conclude from these passages that God brings those events to pass by means of natural processes, by laws he has established and sustains (see Jer. 33:25) rather than by special, miraculous acts. Scripture alone doesn't always make clear which interpretation is correct. The Bible was written long before modern science studied these questions systematically. So it's not surprising that the authors of Scripture were very interested in proclaiming that it is God—the God of Abraham and

Sarah, Isaac and Rebekah—who sustains and controls the natural world, but they were not particularly interested in explaining *how* God does it.

Today, however, the question of *how* does interest us. We are curious about how God causes day and night and makes plants grow, so we systematically study God's world to figure it out. What we learn from these studies can settle questions about how to interpret some particular passages of the Bible. Science does not overthrow the idea that God governs sunrise and sunset, rainfall and harvest. It does, however, inform us that God can govern these things using regular, understandable processes such as the earth's rotation, evaporation and condensation, and the processes of biology and biochemistry. This information, this life experience, guides us in choosing between two acceptable interpretations of the Bible. Science shows us that God governs these things without the need for special, miraculous action.

Rather than placing theology over science or science over theology, remember that God is sovereign over both. The Holy Spirit can guide us to new wisdom and understanding of both. If God uses Scripture to teach something about the natural world, then Christians must listen. If God uses our experiences, including facts learned from science, to improve our understanding of Scripture, then Christians must listen. Science should not cause us to throw out part of the Bible or to interpret it in a way that conflicts with the rest of Scripture. On the other hand, if a passage can be interpreted in several ways, all of which are consistent with the rest of the Bible, then God might use science to help us reach a better understanding of that passage. God created the world, and God inspired Scripture. Our goal should be to listen to what God is telling us from both sources.

Getting BEYOND Controversy

The Apostles' Creed begins "I believe in God, the Father almighty, creator of heaven and earth." It does not go on to specify what we must believe about *how* God created. It is important for every Christian to believe *that* God created the whole universe.

Establishing a firmly held belief about *how* God created the world is not fundamental to our saving relationship with him. Yet because God has given us curious minds and an amazing world, many of us want to learn more about the *how*, and some of us make it our vocation.

In spite of the ranges of interest in and opinions on origins within the body of believers, we feel that every Christian should

▶ take both God's Word and God's world seriously. Simply setting aside one in favor of the other would mean ignoring part of God's revelation.

▶ avoid slandering each other. Whenever one person makes claims about the motives of another person with whom they disagree, that person can become guilty of bearing false witness. For example, some claim that Christians who believe in an old earth have diluted their faith in order to earn respectability in the eyes of secular scientists and that they are on a slippery slope that will lead them to throwing out the whole Bible. Others claim that Christians who believe in a young earth are proudly stubborn and anti-intellectual and that they have made an idol of their own particular interpretation of the Bible. Such claims might be true in a few extreme cases, but they do not explain what truly motivates most Christians to hold their particular beliefs on origins.

▶ avoid setting up unnecessary stumbling blocks to the gospel. Saint Augustine, who lived many centuries before the modern debates about origins, evidently faced a similar problem in his day. He wrote,

> Usually, even a non-Christian knows something about the earth . . . and this knowledge he holds to as being certain from reason and experience. Now, it is a disgraceful and dangerous thing for an infidel to hear a Christian, presumably giving the meaning of Holy Scripture, talking nonsense on these topics; and we should take all means to prevent such an embarrassing situation, in which people show up vast ignorance in a Christian and laugh it to scorn. The shame is not so

much that an ignorant individual is derided, but that people outside the household of the faith think our sacred writers held such opinions, and, to the great loss of those for whose salvation we toil, the writers of our Scripture are criticized and rejected as unlearned men.

—Augustine (A.D. 354-430),
The Literal Meaning of Genesis.

▶ encourage Christian children to pursue science, not fear it. Some young Christians have been blessed by God with an interest in and ability to do science, but they have been warned that scientists make things up to try to disprove God and that their own faith will be at risk if they study science. Instead these young Christians should be encouraged to respond to God's call to a career in science and the task of studying God's handiwork.

Finally a Christian's primary response to God's world should not be debate but an overflowing of praise and worship of the Creator. We see this in Psalm 29, where the biblical author looks at God's world and responds with praise. Imagine the psalmist, perhaps sitting on a hill in northern Israel or standing on the shores of Lebanon, watching a storm move over the land north of Israel. He looks out on the Mediterranean Sea, watches the clouds building, and hears distant thunder. A huge thunderstorm blows onto shore with gusts of wind in the forests of Lebanon, thunder echoing off Mount Hermon (also called Sirion), and lightning cracking over the wilderness of Kadesh. As the storm moves off, the wind dies and the waters calm, and the poet writes,

Ascribe to the LORD, you heavenly beings,
ascribe to the LORD glory and strength.
Ascribe to the LORD the glory due his name;
worship the LORD in the splendor of his holiness.

The voice of the LORD is over the waters;
    the God of glory thunders,
    the LORD thunders over the mighty waters.
The voice of the LORD is powerful;
    the voice of the LORD is majestic.
The voice of the LORD breaks the cedars;
    the LORD breaks in pieces the cedars of Lebanon.
He makes Lebanon skip like a calf,
    Sirion like a young wild ox.
The voice of the LORD strikes
    with flashes of lightning.
The voice of the LORD shakes the desert;
    the LORD shakes the Desert of Kadesh.
The voice of the LORD twists the oaks
    and strips the forests bare.
And in his temple all cry, "Glory!"
The LORD sits enthroned over the flood;
    the LORD is enthroned as King forever.
The LORD gives strength to his people;
    the LORD blesses his people with peace.

Nothing in the psalm talks about *how* a thunderstorm works; there is nothing about meteorology, precipitation, cold fronts, or electrical discharges in the atmosphere. No, the psalm is all about *who*. The storm is under the Lord's control. Every aspect proclaims his authority and glory. The poet artfully uses the human experience of being caught outside in a storm—the gusts of wind, the peals of thunder, the flashes of lightning—to declare God's glory and power. The poem is also a lesson for God's people about the idol Baal, a Canaanite god believed to have been the divine power behind thunderstorms. Not so, the psalmist proudly proclaims; Israel's God, the Lord, is in control. When seen through the eyes of faith, natural wonders like thunderstorms cause us also to cry "Glory!" In a thunderstorm we *experience power*—and sense our own powerlessness. Nature stretches our imagination of what true power might be, so that we have a deeper understanding of the almighty power of God.

QUESTIONS FOR REFLECTION AND DISCUSSION

1. In the Introduction (pp. 11-21) we've shared a bit about our own life experiences and how they've impacted our interests and beliefs about God's world and Word. What have been your experiences around the issues of creation, evolution, and design? What motivates you to learn more about how God created the world?
2. How did you feel about science when you were in school? Did you love it? Fear it? Were you bored by it?
3. Have you heard two or more Christians disagreeing with each other about origins? What were their positions? Were their disagreements cordial or contentious?
4. What other issues can you think of on which all Christians agree on certain basic principles but disagree on particulars?
5. In what situations have you experienced Christians bearing false witness against the motives of other Christians on the topic of origins?
6. The authors ask whether it's ever appropriate for Christians to allow what we learn from the study of creation to affect how we interpret Scripture and vice versa. How do you answer these questions?
7. What are some ways in which the Holy Spirit can guide us into a better understanding of Scripture? What are some ways the Holy Spirit can keep us from unsound interpretations of Scripture?

ADDITIONAL RESOURCES

Educational resources:

Leunk, Thea. *Fossils and Faith.* Grand Rapids, Mich.: Faith Alive Christian Resources, 2005.

This four-week course, designed for high school youth, provides an overview of multiple views on origins that adults will also find helpful. View a sample copy at www.faithalive resources.org.

Vogel, Jane. *Walk With Me Year 3, Unit 5: Discover Creation and Science.* Grand Rapids, Mich.: Faith Alive Christian Resources, 2006.

This unit introduces middle school students to origins issues, such as how the Bible and science answer different sorts of questions.

More on positions of various denominations and Christian organizations:

Barry, A. L. "What about Creation and Evolution?" www.lcms.org/ graphics/assets/media/LCMS/wa_creation-evolution.pdf. Barry discusses the position of the Lutheran Church Missouri Synod that favors young-earth creationism.

Dembski, B., and K. Miller, P. Nelson, B. Newman, D. Wilcox. "Commission on Creation, American Scientific Affiliation," 2000. www.asa3.org/ASA/topics/Evolution/commission_on_ creation.html. "The American Scientific Affiliation is a fellowship of men and women in science and disciplines that relate to science who share a common fidelity to the Word of God and a commitment to integrity in the practice of science." Recognizing the diversity of views among Christians, they have deliberately decided not to take a stand endorsing one particular view of origins but have drafted a statement that recognizes general areas of agreement and specific areas of disagreement among Christians.

National Center for Science Education "Statements from Religious Organizations." www.ncseweb.org/resources/articles/7445_ statements_from_religious_org_12_19_2002.asp. This site gives position statements of several religious organizations that favor an old-earth position and the theory of evolution.

National Public Radio, *Taking Issue.* "Evolution and Religious Faith," 2005. www.npr.org/takingissue/20050803_takingissue_ origins.html. Statements on evolution from leaders of three Christian denominations, as well as Muslim and Jewish leaders.

CHAPTER 2

WORLDVIEWS AND SCIENCE

The fact that you're now into the second chapter of this book indicates that you care about the topic of origins. Why? Why do you want to know more about how God created the world? What deep questions do you hope this study will answer? The conclusions you draw about origins from your personal study or small group discussion will depend on your deeply held beliefs about God and your place in the world. We might call this your *worldview*.

Just what is a worldview? A worldview, or world-and-life view, is often defined as a belief system that a person uses to answer the big questions of life. These questions include the origin of the universe and of humanity, the purpose of human existence, the existence of God, and how one should relate to God. In this context atheism is not the absence of religion. Rather it is a belief system that answers these questions differently than does a God-centered belief system. A person's worldview could be Christian, Judaic, Islamic, Hindu, Buddhist, animist, agnostic, relativist, or atheistic. Within each of these categories a range of worldviews is possible—not all atheists, for example, have exactly the same worldview. Neither do all Christians have exactly the same worldview, although in general Christians share similar worldviews.

> We'll call a worldview a *Christian worldview* if the
> answers it gives to the big questions of life generally
> conform to traditional Christian theology.

In this chapter we'll first consider how these various worldviews make a difference in science and how scientists of different worldviews are able to work together. Then we'll discuss how
those with a Christian worldview understand God's governance of

▶ the natural events that scientists can explain.

▶ the natural events that scientists are still studying but cannot
   yet explain.

▶ supernatural miracles.

▶ apparently random events.

WORLDVIEWS HELD BY SCIENTISTS TODAY

Scientists today hold a variety of worldviews. A significant number of scientists are Christians, including leaders in every major
scientific field. Other scientists follow other world religions,
including Judaism, Islam, Hinduism, and Buddhism. The most
common worldview among scientists today is probably *relativism*. Relativism says that in terms of morality and religion each
person can believe whatever he or she wants to because there
is no absolute truth about such questions and no way to decide
among different beliefs. *Agnosticism*, another common view, says
that the answers to the big questions are unknowable and may
not be that important. Less common among scientists we know,
but more frequently seen in popular books on science, is the
worldview of *reductive atheism*. This view posits that the natural
world is all there is, God does not exist, religion is mere superstition, and truth is only what can be proved logically or experimen-

tally. (In this chapter we will use the term "atheist worldview" to refer to reductive atheism in particular.)

We've seen repeatedly that scientists with very different worldviews can work together comfortably on a professional level. They collaborate on experiments, share theories, listen to each other, and reach agreement on scientific results. How can scientists who have such fundamentally different worldviews so often come to the same scientific conclusions?

Some people have suggested that science, by its very nature, is independent of worldview. Good scientists, they say, are simply objective; when they enter the lab they set aside all prejudice and beliefs. But the history of science shows that worldview beliefs frequently do influence scientific choices. Besides, the idea that there is such a thing as *objective truth* is, in itself, a worldview belief.

Other people have suggested that scientists set aside their own *personal* worldviews and temporarily adopt a *professional scientist* worldview in which supernatural beings like God do not exist. The idea here is that when scientists try to understand natural laws they do not allow supernatural causes to be part of the discussion. (For example, when scientists say, "The law of gravity causes apples to fall to the ground," they make no mention of God and are therefore acting as though God doesn't exist.) From a Reformed Christian perspective we find this description of science to be both inaccurate and repulsive. We believe that faith is a part of everything we do, not something that is set aside during weekday science experiments. But if this view of science is inaccurate, we still face the same questions: If an atheist and a Christian work together on a scientific experiment and reach the same conclusion, does this mean that the Christian has abandoned her faith? Or that the atheist has abandoned his worldview? Or does it mean neither? An answer to these questions lies not in considering the ways in which those worldviews differ but rather in considering what they have in common.

WORLDVIEW BELIEFS NECESSARY FOR SCIENCE

All scientists, regardless of their particular worldview, hold certain philosophical beliefs foundational for doing science. Some of these are listed in the left-hand column on the chart on page 43. These fundamental beliefs cannot be proved from science itself. The fact that science actually works lends support to these beliefs, but the beliefs themselves come from outside of science, perhaps from culture, or religion, or simply the scientist's personal choice. Today these beliefs may seem obvious, but for most of human history people did not hold to all of them. Animists, who believe that gods inhabit many aspects of the physical world, would have very different views of cause and effect and the regularity of nature. Plato and Aristotle developed logical and beautiful theories about the workings of the natural world, but they got some of the answers very wrong because they did not place enough priority on doing experimental tests. Even today people who follow astrology or *new age* beliefs would disagree with some of the beliefs listed in the left-hand column of that chart.

Consider some Christian theological beliefs that come from biblical teachings about God and the world. We've listed several in the right-hand column on the chart on page 43. Notice how each Christian belief on the right naturally gives rise to the worldview belief on the left. For a Christian, biblical teachings about God and the natural world provide ample support and motivation for doing science and a basis for understanding why science is so successful. Christians doing science are not acting as though God doesn't exist. Rather, they are acting on their belief that there *is* a God—not a capricious God, but the God of the Bible who made an orderly world and still governs it in an orderly fashion.

This also helps us understand why Christians who are professional scientists usually come to the same *scientific* conclusions as scientists with other worldviews. Although scientists with other worldviews do not share with Christians the beliefs about God and the meaning of human life listed in the right-hand column

Worldview Beliefs Needed for Science

Humans have the ability to study nature and to understand, at least in part, how it functions.

Christian Beliefs

Humans are God's imagebearers in this world (Gen. 1:27). Thanks to the abilities that God has given us, we can understand, at least in part, how the world works.

Events in the natural world work by natural cause and effect. For example, a tree falls because the wind exerts a force on it, not because it wanted to fall, nor because a forest god made it fall, nor because it simply was fated to fall.

There are no nature spirits, no capricious gods, no fate. There is only one God (Deut. 6:4) who created and rules the world (Gen. 1) in a faithful, consistent manner (Ps. 119:89-90).

Natural phenomena are repeatable; they are regular across space and time. Scientists will find the same experimental result in laboratories all over the earth, and will find the same result today as they found last week. This consistency allows the phenomena to be studied using logic and mathematics.

God has established natural laws (Jer. 33:19-26) and faithful covenants (Gen. 8:22) with the physical universe. So we are not surprised to discover that nature typically operates with regular, repeatable, universal patterns.

Observations and experiments are necessary to build and test scientific models that correctly describe natural phenomena. Logic and deduction alone are not sufficient to build an accurate understanding of the natural world.

God was free to create the world in many ways. Humans are limited and sinful. We are unable to understand God's ways completely (Job 38). So our scientific models based on logic and deduction must also be tested by careful experimentation and observation, comparing them to what God has actually made.

Science is a worthwhile use of human time and resources.

Studying nature is worth doing because we are studying the very handiwork of God (Ps. 19:1). God has called us to study his creation (Gen. 2:19-20; Prov. 25:2) and to be stewards of it (Gen. 1:28-29; Ps. 8:5-8).

of the chart, they do share the beliefs in the left-hand column. Sharing that common *subset* of beliefs with Christians means that they can work together as professional scientists and reach consensus. This would not have surprised John Calvin, a theologian and church reformer from the 1500s, who wrote, "All truth is from God, and consequently if wicked men have said anything that is true and just, we ought not to reject it, for it has come from God" (*Calvin's Commentaries* on Titus 1:12). Calvin also wrote,

If the Lord has willed that we [Christians] be helped in physics, dialectic, mathematics, and other like disciplines, by the work and ministry of the ungodly, let us use this assistance. For if we neglect God's gift freely offered in these arts, we ought to suffer just punishment for our sloth.

—*Institutes of the Christian Religion,* 2.2.16.

Many scientists throughout the centuries have seen their belief in God as completely compatible with their scientific work. We've included the testimonies of some scientists in our own fields of physics and astronomy on our website (www.faithaliveresources.org/origins). Look for the article "Scientists of Faith."

WORLDVIEWS AND GOD'S GOVERNANCE

God's Governance of Explainable Natural Events

When scientists observe regular patterns in nature, such as the cycle of the seasons or the growth of new grass every spring, they try to understand how they work. In some cases the patterns are so universal that scientists call them *natural laws*. The law of gravity is a prime example. Every single object scientists have observed in the universe obeys the law of gravity. Gravity doesn't change from day to day. In fact, its strength is so predictable that scientists can describe it with a mathematical equation. Because gravity is so regular and reliable, scientists sometimes say "The law of gravity governs the solar system."

But Scripture tells us that this isn't the whole story. Natural laws don't govern; *God* governs. God speaks of his "covenant with day and night and the fixed laws of heaven and earth" (Jer. 33:25). The regular patterns of day and night, summer and winter, and other fixed laws of nature were established by God's design. The fact that these patterns are so regular and understandable is a gift from God, without which we would not be able to understand the world.

Participants in clashes between religion and science often forget this. Those in the religion camp see every new scientific discovery as a challenge to God. They think that if some aspect of nature is understood by science, then God isn't needed or is less involved. But a god who becomes unnecessary as soon as humans find a scientific explanation for how the world works is not the God of the Bible! The Bible clearly teaches that natural, ordinary events are governed by God. God is active not only in dramatic supernatural events but also in the ordinary changes of the seasons. This view of God is displayed in Psalm 104:19-24:

He made the moon to mark the seasons,
    and the sun knows when to go down.
You bring darkness, it becomes night,
    and all the beasts of the forest prowl.
The lions roar for their prey
    and seek their food from God.
The sun rises, and they steal away;
    they return and lie down in their dens.
Then people go out to their work,
    to their labor until evening.
How many are your works, LORD!
    In wisdom you made them all;
    the earth is full of your creatures.

This psalm describes events both as natural events and as divine actions. The sun goes down (natural event), and God brings night (divine action). The lions hunt for prey (natural event), and they seek their food from God (divine providence). The Bible clearly proclaims that God is fully in charge of natural events. A scientific explanation of a natural event does not replace God, and it doesn't mean that God is absent. Donald MacKay, a Christian and a physicist, described God's providence in his book *The Open Mind and Other Essays* (1988): "The continuing existence of our world is not something to be taken for granted. Rather it hangs moment by moment on the continuance of the upholding word of power of its Creator." In this view of providence, God sustains natural laws and the very existence of atoms and light waves. All matter and energy, all space and time, are continually and actively maintained by God.

MacKay called this view dynamic stability because although these things look stable and unchanging, they are maintained through an ongoing, dynamic process. He compared it to a computer game. Suppose you are playing a computer game that simulates a pool table. As the player, you control the cue stick and shoot the balls, and the computer moves the balls around on the screen. The balls follow all the rules you'd expect: they go faster if you hit them harder, they roll in straight lines, and they bounce off the sidewalls or off each other. On the computer screen the balls appear solid and self-existent, and they follow regular, repeatable patterns of behavior (the "natural laws" of the computer game). But that doesn't mean that the electronic pool balls will continue to exist when you pull the plug on the computer! The pool balls are not solid and stable on their own. Rather, the computer continuously sends signals to the screen to update and maintain the image. The laws that govern the motion of the balls aren't stable on their own either; if a glitch occurs in the program, the balls will freeze and no longer follow the rules.

Similarly, MacKay says, the matter, energy, and laws of nature of this universe are not of themselves intrinsically self-existing and stable; nor did God simply start them off and then leave them alone. They owe their continued existence and apparent stability to the fact that God continually wills it.

Of course, no one can fully understand how God governs the natural world. As scientists do experiments and make hypotheses, they discover some aspects of the regular, repeatable patterns that God placed in nature. Even a partial understanding of those patterns, such as Isaac Newton's understanding of gravity, can be enough to describe the natural world so accurately and reliably that scientists call their understanding a natural law. That does not make natural law identical to how God actually governs; it's possible that details and subtleties are not yet understood by scientists. In fact, new experiments and theories often change how scientists understand a pattern in nature, as when Albert Einstein developed a new understanding of gravity (general relativity) that makes even more accurate predictions than Newton's law of gravity. Because scientists can never be sure they currently understand every aspect of a pattern in nature, it is inaccurate to claim that they have discovered the natural laws God uses to govern nature. It is more accurate to say "God usually governs nature in regular patterns, and to the extent that we understand these patterns, we call them natural laws." (In this book we will sometimes use phrases like "God governs through natural laws" as shorthand for the second statement.)

Despite the limitations of our understanding, scientists over time do learn more and more about nature. As they form new models and make new predictions that are tested by experiments, they come to a better understanding of the patterns in nature, and thus a better understanding of how God governs the world. Christians who are scientists can find this process of scientific discovery to be a worshipful and awe-inspiring experience.

Christians can celebrate rather than fear advances in scientific knowledge. When scientists explain some part of the natural world in terms of natural laws, this does not remove God from the picture. Rather, science helps us to partially understand the patterns of God's governance.

The question of how God governs natural events was debated in the early days of science when Newton and LaPlace studied planetary orbits. To learn more, click on "How Does God Keep Planetary Orbits Stable?" on our website (www.faithaliveresources.org/origins).

God's Governance of Unexplainable Natural Events

God's world surprises us with unexplainable natural events. Consider the annual migration of monarch butterflies. A butterfly crawls out of its chrysalis in the north, flies hundreds of miles, and somehow finds the home of its ancestors in Mexico, even though it has never been there before. Some Christians have responded to this marvel as follows:

> Entomologists have studied this mystery for years, but have no answers as to how it happens.... If there were no other evidence of design in God's Creation than caterpillars and butterflies, this alone would be enough to show the fact of His design in His Creation.
>
> —Project Creation, "Butterflies: The Miracle of Metamorphosis" www.projectcreation.org/creation_spotlight/spotlight_detail.php?PRKey=32.

As Christians we believe that God designed and created the whole world, including caterpillars and butterflies. We praise God for the beauty of monarchs, for the intricacies of their life cycle, and for their amazing migration. But can we use this as yet unexplained mystery to *prove* that God is Creator and Designer?

Such arguments have failed in the past. Given the amount of research that is happening today, scientists will almost certainly figure out the natural causes of butterfly migration and navigation. If this happens, what is left of the argument? Where is God? Arguments like this focus on the gaps in scientific knowledge—the things scientists don't yet understand. The gaps in our scientific understanding are set forth as the best places (or only places) to see God's hand at work. But scientific knowledge keeps growing, and the gaps keep shrinking. If God is only a "god of the gaps," then God would shrink as scientific knowledge grows. In fact, many atheists and agnostics believe that the explosion of scientific knowledge over the centuries is evidence that the idea of *God* is irrelevant. Christians play into their hands when they make arguments for God's existence based on gaps in our scientific knowledge.

A better approach is to acknowledge and proclaim God's design and creative hand in both the things science cannot explain *and* the things it can. God governs the regular functioning of the natural world, whether or not science understands it yet. This approach bears truer witness to who God really is and will not become irrelevant as science advances. If scientists come to understand monarch migration, we can still proclaim God's faithful governance and providential care of butterflies. As more and more natural phenomena are understood by scientists, our understanding of *how* God governs will increase. But God's design in nature will not change, nor will nature's utter reliance on God's sustaining hand. God does not shrink as science advances.

When scientists finally figure out the natural process by which butterflies migrate, it will be amazing, maybe involving some complex genetic code or some unique ultraviolet sensors in the butterfly's eyes. Whatever it is, won't that knowledge *increase* our praise of God as the designer of that natural process? In the words of Proverbs 25:2: "To search out a matter is the glory of kings." God calls on humans to use their God-given gifts to work out mysteries, to rejoice in following the clues and figuring out the mystery of *how* God works through natural processes.

God's Governance of Supernatural Miracles

Some people believe that science and religion clash on the topic of miracles. Christians believe that God can do and has done miracles. While some biblical miracles, such as the earthquake in Numbers 16 and the plague of locusts in Egypt, do not directly contradict the laws of nature, other miracles are obviously supernatural. Jesus turned water into wine, healed the sick, and rose from the dead. But science says that according to the laws of nature—the way the natural world ordinarily acts—it is impossible for water suddenly to turn into wine or for a person to rise from the dead.

That's exactly the point. These things could *not* happen in the ordinary course of events. Science really doesn't have anything new to say here that wasn't known in biblical times. The pre-scientific people of the Bible certainly could tell the difference between ordinary, natural events and supernatural ones. Water does not spontaneously turn into wine at most wedding banquets! Many of the miracles of Moses, Elijah, and Jesus were intended to be obviously supernatural, to demonstrate the presence and power of God to the people.

So does science challenge our belief in miracles? If someone believes that natural laws invariably govern nature, then the answer is yes. In that case, natural laws are always obeyed, without exception, and that would prevent resurrection from the dead. But Christians believe that *God*, not natural laws, governs nature. God typically works through natural laws to sustain the regular patterns of our world, but nature is not locked into those patterns. When it suits God's kingdom purposes, God works outside the ordinary patterns of governance and we see a miracle. The rest of the time God works through those ordinary patterns and we see events that we can describe scientifically, using natural laws.

By their very nature supernatural miracles cannot be explained by scientific study. They are unique events that do not follow the regular patterns seen elsewhere in nature. Science can study only natural cause and effect, but miracles do not have ordinary natural causes. At most, science can say "When natural laws operate ordinarily, water does not spontaneously change into wine." Otherwise, science has nothing to say about whether or not miracles are possible.

God's Governance of Random Events

We've often heard Christians say "Life on earth is so amazing, it couldn't have come about by chance; it must have been made by God." And we've heard non-Christians say "Life is the result of random processes; no God is involved." We disagree with both statements. To explain, we first need to tell you what scientists mean by the word *random* (or its synonym *chance*).

When scientists say that something is *random*, they mean that the outcome is unpredictable. Consider the roll of a pair of dice. Scientists can calculate the probability that the roll will yield a five or an eleven, but they can't predict what any *particular* roll will turn out to be. It's not that some mysterious force is at work making the dice roll differently each time. Rather, each time the dice are rolled they follow exactly the same well-understood natural laws of gravity and motion. The dice land differently each time because of how they bounce and spin. If the dice are tossed even slightly differently from one time to the next, that slight difference is magnified by each bounce, and after several bounces the final outcome is completely changed. The system is scientifically random because the outcome is unpredictable.

The weather is also scientifically random. Meteorologists understand a lot about warm fronts and cold fronts and the general behavior of the atmosphere at different temperatures and pressures. They understand the natural causes of rain. A meteorologist might forecast a 20 percent chance of rain. Meteorologists can even make good ballpark forecasts about the temperatures and cloudiness over the next several days. But, like the rolling of dice, the motions of the atmosphere are too complicated for precise predictions. Because the system is complicated, meteorologists can calculate probabilities, such as the average cloudiness during the afternoon or the chance that it will rain, but they cannot calculate the exact outcome, such as whether a cloud will be over your house at two o'clock tomorrow afternoon.

This scientific use of the term *random* (or *chance*) is entirely compatible with a biblical picture of God's governance. Many Bible passages describe God working through events that appear to be random from a human perspective. Consider Proverbs 16:33: "The lot is cast into the lap, but its every decision is from the LORD." Centuries before modern science, ancient peoples knew that some events, like the casting of lots and the weather, are unpredictable. Yet this passage from Proverbs, along with other Bible passages, proclaims God's sovereignty over events that scientists today describe using probabilities. Events that appear random from a human perspective pose no problem to a biblical understanding of God's providence. God can control *"chance"* events that are unpredictable to us.

Someone may say, "Life came about by chance, not by God." In statements like this, the word *chance* is being used with a very different meaning from the way scientists use it. Here it is being used in a philosophical sense to mean *a lack of cause and purpose.* In statements like this, *chance* functions almost like a god that is set up in opposition to the God of the Bible.

Christians believe that nothing is outside God's governance, including scientifically random events. God can use these events to accomplish his purposes, and God's will gives them meaning. When Christians say "Life on earth is so amazing, it couldn't have come about by chance," they could simply substitute the words "without God" for the words "by chance," and the sentence would convey their intended meaning.

Many English words have more than one definition. The problem comes when the two meanings of *chance* (scientific and philosophical) are confused. This usually happens when an event that is scientifically unpredictable is assumed to be lacking meaning and purpose. But the question of meaning and purpose is a religious one, and science cannot answer it. Science can only say that the particular details of a rainstorm are unpredictable, not that a particular storm is God's answer to the prayers of a drought-stricken farming community or even whether all storms are governed by God. Science does nothing to challenge our biblical belief that God governs events that appear random from a human perspective.

Proverbs 16:33 indicates that God can select particular outcomes in systems that are scientifically random. God has designed some phenomena (like casting lots or genetic mutations) to happen in scientifically unpredictable ways. But in any, or perhaps every, particular random event, God could choose to select one particular outcome. God might do this subtly, controlling particular events in ways that are significant but that we cannot detect scientifically because the overall set of events has a random distribution. God could also choose to do this dramatically upon occasion, choosing an outcome that is scientifically possible but extremely improbable, something that might even appear miraculous to us.

Does God directly control each random event, or does God let created systems explore options within bounds that he sets? For two possible answers to this question, visit our website (www.faithaliveresources.org/origins) and click on the article "God's Governance: Two Views."

In the introduction to this chapter we raised the question "Who governs?" The answer is God. According to our Christian worldview,

▶ God governs the natural events that scientists can explain, like the cycle of seasons and plant growth.

▶ God governs natural, regular events that scientists are still studying but can't yet explain, like the migration of monarch butterflies.

▶ God governs supernatural miracles that science in principle cannot explain, like the resurrection and water turning into wine.

▶ God governs random events in which scientists can't predict the outcome, like the roll of dice and the weather.

To God be the glory!

QUESTIONS FOR REFLECTION AND DISCUSSION

1. People sometimes say that it takes just as much faith to believe in science as it does to believe in religion. Is that a helpful comparison? Why or why not? Do most scientists believe some things even though they can't prove them? What sorts of evidence do you have for your religious beliefs?

2. Do you believe that all truth comes from God? How do you explain the fact that many discoveries are made by people who don't believe in God?

3. How would you respond to the claim that God isn't necessary because science can explain how the world works?

4. How would you respond to the claim that God must exist because scientists can't explain something in the natural world?

5. Which do you think is a better analogy for God performing a miracle: Is it like an airplane pilot who takes the plane off "autopilot" to perform a maneuver and then puts it back on "autopilot"? Or is it like a piano player who plays notes in a beautiful pattern and occasionally plays a few notes in a dramatically different way? Can you think of a better analogy?

6. Steven Weinberg, a Nobel prize-winning physicist and an atheist, said,

   I think we should be very careful not to give an impression to the public that somehow our scientific work is converging with religious work into a synthesis. It's not. I don't want a constructive dialogue. I don't want to do anything to reconcile science and religion. I think it's very good that they remain at odds with one another (quoted by Jeremy Beer in "Dad, I'm an Atheist," *Re:Generation Quarterly*, July 1, 2000).

   Why do you think Weinberg is so opposed to reconciliation of science and religion? In what ways could a Christian respond to Weinberg or others like him?

ADDITIONAL RESOURCES

Ecklund, Elaine Howard. *Science vs. Religion: What Scientists Really Think.* Oxford: Oxford University Press, 2010.

Grinnell, Frederick. *Everyday Practice of Science: Where Intuition and Passion Meet Objectivity and Logic.* Oxford: Oxford University Press, 2009. A cell biologist describes what doing science is really like.

Haarsma, Loren. "Chance from a Theistic Perspective," *The Talk Origins Archive,* 1996; www.talkorigins.org/faqs/chance/chance-theistic.html.

_____. "Does Science Exclude God? Natural Law, Chance, Miracles, and Scientific Practice," *Perspectives on an Evolving Creation.* Keith B. Miller, ed. Grand Rapids, Mich.: Wm. B. Eerdmans Publishing Company, 2003.

Hooykaas, R. *Religion and the Rise of Modern Science.* Scottish Academic Press and Chatto & Windus, 1972.

MacKay, Donald. *Science, Chance and Providence.* Oxford: Oxford University Press, 1978.

Murphy, George L. *Toward a Christian View of a Scientific World.* Lima, Ohio: CSS Publishing Company, 2001.

Polkinghorne, John. *The Faith of a Physicist.* Princeton, N.J.: Princeton University, 1994.

_____. *Science and Providence.* Boston: Shambhala Publications, 1989.

CHAPTER 3

SCIENCE: A PROCESS FOR STUDYING GOD'S WORLD

As we write this, it's March in Michigan—the dreary end of a long winter. The snowfall that was so delightful in November now feels oppressive. Even when the snow melts, the skies remain gray, as gloomy as the dead grass and bare tree trunks. Intellectually, we know that spring will come and our yard will become green again, but emotionally it seems to take forever! At these times it can be reassuring to recall God's promise in Genesis 8:22: "As long as the earth endures, seedtime and harvest, cold and heat, summer and winter, day and night will never cease." God still faithfully governs the seasons and will bring warmer weather in due time. God's Word assures his people about the *who* of creation. We discussed this part of our worldview in chapter 2.

Since we're scientists, we also like to remind each other about *how* God will bring the spring weather. God is governing the earth's orbit around the sun using the natural laws of momentum and gravity. As the weeks pass and the earth moves around the sun, the tilt of the earth's axis yields longer days and more direct sunlight in the northern hemisphere, including the grass in our yard. These are things learned by carefully studying God's world; they are part of a body of scientific knowledge about God's creation.

But science is more than a body of knowledge. Science is better described as a *process* by which people gain that knowledge. In this chapter we'll describe three methods used in the process of gaining scientific knowledge: *experimental, observational,* and *historical.*

Since this book deals extensively with historical science, we'll consider the similarities and differences among these three categories and then discuss how historical science reveals how God's world behaved in the past.

THREE METHODS OF SCIENTIFIC INVESTIGATION

All three methods of scientific investigation include building *models* (also called *hypotheses* or *theories*). A scientist's model is a set of ideas about the physical cause-and-effect behavior of the natural world. It attempts to explain the results of past experiments and observations and makes predictions about what future experiments and observations will reveal.

Experimental Science

Controlled experiments are one of the most important tools of science. Here is an example of a controlled experiment:

In a laboratory, a biologist carefully counts out seeds and plants them in two containers of soil, one container held at a cooler temperature and one at a warmer temperature. She keeps them watered for several days and counts the seedlings when they appear. She calculates the fraction of seeds that actually sprouted (the germination rate) and notes that it is higher in warmer soil than in cool soil. She decides on a model that seeds sprout better in warm soil than cool soil. Based on this model, she predicts that in very cold soil no seeds will germinate and in very hot soil all seeds will germinate. To test the model and the prediction, she throws out the first batch of plants and runs the experiments again, this time using several containers of soil at several different temperatures. The results for cool and warm soil agree with the first experiment, confirming those results. No seeds sprout in the coldest soil, confirming that prediction. But in the hottest soil, no seeds sprouted at all! She modifies her model: germination rate increases with soil temperature, but at the hottest temperatures the seeds are "cooked" before they can sprout.

Experimental science is the primary type of science done in the fields of physics, chemistry, and molecular biology, as well as in certain aspects of ecology and geology. In the laboratory experiments are accessible; the scientist can measure what is happening, monitor the experiment from beginning to end, destroy the products of the experiment, and start over at any time. She can control many variables in the experiment (such as soil temperature) and remove external variables (such as rabbits eating the seedlings). And she can repeat experiments in the lab as necessary to confirm the first results. Basic assumptions, such as the underlying consistency of physical laws, can be tested by repeated experiments. Experimental scientists make testable predictions (such as the germination rate in very hot soil) that can be confirmed or contradicted in future experiments.

Observational Science

Another important tool of science is making careful observations. Sometimes controlled experiments cannot be done because the system under study won't fit in the lab, is too far away, or is too dependent on its environment. In those cases scientists can still make careful observations. This method is also used when there are ethical reasons to avoid a full range of experiments, such as in medicine. Here's an example of observational science:

An ecologist travels to the site where a forest fire occurred the previous year to study how the forest is recovering. He carefully counts all of the plant seedlings in a certain area and notes what types of plants they are. For the next ten years he returns once a year to count the growing plants. He finds that the wildflowers are the first to sprout and grow, but after some years tree seedlings are starting to compete with them. He hypothesizes that wildflowers grow better than tree seedlings in sunny, open spaces, but as the area gets crowded with plants and becomes shadier the tree seedlings do better. He predicts that this same recovery pattern will be observed at other forest fire sites and that the growth of tree seedlings will depend on the amount of shade in the area. He repeats the observations at another fire site where the fire was more widespread. Because fewer trees survived this fire, the site has less shade. He confirms his hypothesis that wildflowers are the first plants to appear but sees far fewer tree seedlings than at the first site. This corresponds with his model that shady conditions are important for the appearance of tree seedlings.

Observational science is commonly done in the fields of meteorology, ecology, medicine, astronomy, and geology. Typically the objects studied in observational science are less accessible than those in experimental science. The ecologist can't sit all year and watch the plants grow, and an astronomer can't travel to a star to measure its temperature. But scientists devise alternate methods to get around these difficulties, such as counting plants periodically or analyzing the light of the star to deduce its temperature. Observational science is not controlled; meteorologists cannot produce a cold front whenever they like, nor do ecologists burn down forests just so they can watch how they recover. Observational science must take nature as it comes. A lack of control makes observational science less repeatable than experimental science. The forest fire can't be repeated whenever the ecologist wants, but fires happen often enough that many sites are available to study. Usually enough examples are available that the consistency of the underlying laws of nature can be tested on several cases. Observational science is a useful partner to experimental science, because observations can be made in many situations where experiments cannot be done. But just like experimental science, observational science makes testable predictions (like the wildflower growth rate after a fire) that can be confirmed or contradicted in observations of other, similar systems.

Historical Science

A third method of scientific investigation is modeling the past behavior of systems, including events that occurred before they could be directly observed. Here's an example:

> An ecologist travels to a remote forest in order to study its history. She first examines a large tree that has recently fallen down in a storm. She takes a thin slice of the trunk back to the laboratory and counts the tree rings. She finds that a particular ring from 131 years ago is extremely thin (indicating drought) and shows evidence of mild fire damage. She hypothesizes that much of the surrounding forest burned down 131 years ago but that this tree survived. Based on the work of her colleague who studies recent forest fires, she makes predictions about the other trees living in the forest: the largest trees will show similar fire damage from 131 years ago; many of the smaller trees will prove to be 120-125 years old, having sprouted 5-10 years after the fire. To test this prediction, she takes core samples of several living trees and looks at their rings. The results confirm her prediction: the older trees all show evidences of fire damage from 131 years ago, and many of the smaller trees are about 120 years old.

Historical science is common in the fields of ecology, climatology, astronomy, cosmology, evolutionary biology, geology, and paleontology. The goal of historical science is to deduce the natural history of systems such as forests, rocks, and planets. Historical science is not directly accessible because no scientists were around at the time to make observations; however, those events are indirectly accessible because of the evidence they have left behind. Like a detective, a historical scientist uses the evidence available today to deduce the history.

Like observational science, historical science is not controlled: scientists cannot go back in time to change the initial event, so they have to work with what actually happened. Historical science investigations can be repeatable when many similar historical situations are available to study (such as the many different trees born after the same forest fire). In some cases, however, the event is not repeated (as in the case of the universe: there is only one universe for cosmologists to study), but scientists can still find evidence that tells them about the natural processes that occurred during that event. Historical science, at its best, is particularly useful for testing whether physical laws remain unchanged over the years, because historical science gathers data related to events that happened over as wide a period of time as possible.

Most important, historical science makes *testable predictions*, just as experimental and observational science do. Scientists routinely study one system (such as one tree or one star cluster), make a model for its history, and then predict what they will find in additional observations. These observations could be of other similar systems, or they could be of the same system but made with different instruments. In either case the observations test the prediction, supporting or contradicting their model for the history of the system.

For another real-life example of historical science, check out the article "K-T Boundary Investigation" on our website (www.faithaliveresources.org/origins).

Crab Supernova Remnant

Even before the invention of the telescope, many cultures around the world made careful observations of the heavens, recorded comets and sunspots, and designed calendars and star charts. In A.D. 1054 Chinese and Arabic astronomers recorded that a bright "guest star" appeared in the sky; they noted the constellation in which it appeared and the date. Native American and other cultures made less detailed records but seem to have observed the same event. The guest star was bright enough to be seen in the daytime for a few weeks; at night it initially outshone all the other stars and took two years to fade away.

Fast-forward to the early 1900s. Western astronomers with modern telescopes but with no knowledge of the Chinese records observe and photograph the Crab Nebula. By comparing photographs taken decades apart, they can see that the nebula is expanding, suggesting that it is the remains of an exploded star. They calculate the rate of expansion and run it backward, concluding that the supernova explosion must have occurred about 900 years earlier.

This is another example of historical science. The astronomers built a model of the past history of the system based on present observations and other scientific information, then used the model to make testable predictions. They began checking historical records to see whether the original burst of light had been noted by any cultures 900 years earlier and found the detailed records of the Chinese and Islamic astronomers. Not only did the date of these ancient records match the calculations of modern astronomers, but the position in the sky agreed as well. This confirmation with written human records shows that historical science is not only possible but can be accurate and reliable.

Photograph by David Malin

This photo shows another supernova explosion that occurred much more recently, on February 24, 1987. The image on the left shows the star before it exploded. The image on the right shows how much brighter it was after the explosion. At its brightest, a supernova can outshine the light of an entire galaxy, brighter than a billion stars! Such a dramatic explosion is one of the ways God reveals himself in the natural world, showing us his incredible power.

All Three Methods Needed

These three styles of science blend into each other. As experiments become more complex and less controlled, they become more like observational science. Observational scientists often do experiments in labs to help them better understand what they are learning. The historical models used in historical science depend on observations of present data, and observations of present behavior are more useful when you have a good model of the past history of the system.

All three methods of gaining scientific knowledge are necessary. Systems such as forest recovery are too large and complex to be brought into the laboratory and must be studied observationally. But laboratory experiments about seed germination are helpful to foresters in understanding which plants are most likely to sprout after a fire. Historical studies of past fires help scientists understand how forests can recover over the long term. Each of these three types of science can make new discoveries that are then tested by the other types of science. These different styles of investigation reinforce and correct each other, leading to a better understanding of the natural world and its history.

HUMANITY'S SIN AND HISTORICAL SCIENCE

Genesis 3 describes humanity's fall into sin, sometimes called "the curse." Verses 14-19 describe consequences for humans, including increased pain in childbirth and painful toil in the field against thorns and thistles. These verses have stimulated a lot of speculation about exactly how much the created world changed at the time of the Fall. Were the fundamental laws of nature radically changed when humanity fell into sin? Or were the effects of the Fall and the curse limited primarily to changes in how human beings relate to God, to each other, and to the natural world?

Scripture does not directly teach that the laws of nature were radically altered by humanity's fall into sin. But some Christians support such an interpretation of Genesis 3:14-19 in light of Romans 8:20-21:

> For the creation was subjected to frustration, not by its
> own choice, but by the will of the one who subjected it,
> in hope that the creation itself will be liberated from
> its bondage to decay and brought into the freedom
> and glory of the children of God.

Other Christians interpret Genesis 3:14-19 to say that the effects of the Fall and the curse are more limited. They argue that the surrounding context of the verses in Romans 8 shows that the passage is about *our* transformation, through Christ, from bondage to sin to becoming children of God. Creation is frustrated now because humanity is not playing its proper role as sinless image-bearers of God and stewards of creation. These Christians also point to Jeremiah 33:16-26 and God's "covenant with day and night and . . . the laws of heaven and earth." God appears to be keeping this covenant today in the same way that he did before the Fall and the curse. Arguing that creation *today* still declares God's glory, they also point to passages such as Psalm 19:1: "The heavens declare the glory of God; the skies proclaim the work of his hands."

So Christians have at least two interpretations of Genesis 3:14-19, each of which has scriptural support. Since this is a question about the natural world, can a study of the natural world help resolve the issue? In this case the answer is yes. Historical sciences like astronomy and geology inform us about how the natural world behaved in the past.

Astronomers have found that the light of a star contains detailed information about the natural laws at work in the star, including laws for gravity, pressure, the behavior of atoms, and the speed of light. When scientists compare the light from our sun, nearby stars, and the most distant stars, they see exactly the same fundamental laws of nature in operation in every case. Because light takes time to travel, the light we see today actually left those distant stars a long time ago. That means that the light we see today shows us what the fundamental laws of nature were like in the distant past. The fact that all distant stars show the same laws of physics as nearby stars is clear evidence that the laws of nature have not radically changed at some point in the past.

Geologists have similar evidence. In their study of rocks they also discover the fundamental laws of nature that were in operation when those rocks were formed. By comparing more recent rocks with the oldest rocks on earth, they find clear evidence that the laws of nature have not radically changed in the past.

We'll discuss the Fall and its effects again in chapters 11 and 12. For now we want to emphasize only one point: whatever the effects of the Fall on humanity, the study of nature tells us that the Fall did not fundamentally alter how atoms and molecules and rocks and stars behave. This means that historical sciences like geology and astronomy really can tell us something about how God has governed creation in the past. It also means that even though fallen humanity cannot comprehend it as we should, "the heavens declare the glory of God" just as much today as when they were first created.

SCIENTIFIC KNOWLEDGE: ONE KIND OF KNOWLEDGE AMONG MANY

Science is a powerful, reliable tool for answering questions about the natural world and its history. (We'll say more about reliability in ch. 4.) But science is *incapable* of answering many other questions. For these we turn to nonscientific methods of understanding the world and other ways of gaining reliable knowledge. For instance, if a police officer wants to determine what happened during a crime, she uses eyewitness reports. If a historian wants to learn about life in the past, he reads historical documents. To judge the trustworthiness of a friend, we don't perform scientific experiments; instead, we decide based on interpersonal experience with that friend. Our aesthetic experience tells us about the quality of a piece of music or art. Prayer and worship help us to better understand the character of God and to experience God's presence. These various methods of gaining knowledge do not involve controlled scientific experiments and mathematical modeling, but they are valid, potentially reliable ways for gaining knowledge in these other areas.

Because science has been so successful, some people claim that it is the one and only way of obtaining reliable knowledge. This idea arises from a worldview of *reductive atheism*, which looks at everything, including human beings, at only the scientific level. A reductionist might express this belief along these lines: "Humans are nothing but chemical machines governed by their genetics and hormonal reactions." With the little phrase "*nothing but*" the reductionist has denied the validity of all other levels of human experience and knowledge. Author Donald MacKay points out that this is like saying that a Shakespearean sonnet is nothing but ink on paper (see *The Open Mind and Other Essays*, 1988). But of course a sonnet is so much more than ink on paper! It is letters and words, rhyme and meter, allusion and metaphor, and expressions of love or pathos. We know these things about a sonnet not from science but from our understanding of language, literature, and life experience. All of these types of knowledge working together give us the best understanding of the poem.

Here's a trick you can use if you encounter a "nothing but" argument: replace *"nothing but"* with the words *not only*. For example, "Yes, human bodies are made of chemicals, but human beings are not only chemical machines. They are also complex biological systems, relating to one another in social groups, forming culture and artwork, and relating to their Creator God."

Scientific knowledge, then, supplements but does not replace important truths we learn from history, Scripture, personal experience, and culture. Science is one of many gifts God has given humanity.

QUESTIONS FOR REFLECTION AND DISCUSSION

1. Is it possible to study a process scientifically if you weren't there to observe it from beginning to end? In what ways is historical science like a detective putting together clues at a crime scene? For example, do detectives make models? Do they make predictions and test them?
2. Why is it important for Christians to pursue scientific careers? What impact could they have on scientific knowledge? On the scientific community? On the church?

ADDITIONAL RESOURCES

Hearn, Walter. *Being a Christian in Science.* Downers Grove, Ill.: InterVarsity Press, 1987.

MacKay, Donald. *The Open Mind and Other Essays.* Leicester, England: InterVarsity Press, 1988.

Ratzsch, Del. *Science & Its Limits: The Natural Sciences in Christian Perspective.* Downers Grove, Ill.: InterVarsity Press, 2000.

CHAPTER 4

GOD'S WORD AND GOD'S WORLD IN CONFLICT?

I n 1633 Galileo Galilei stood on trial before the Inquisition in Rome. Galileo had argued that the Earth and all the other planets traveled through space in circular orbits around a stationary sun. While this truth is now taught in every elementary school textbook, at that time it directly contradicted church teaching. The Vatican's position was that the Earth was fixed in space and that the sun moved in orbit about the Earth. On Galileo's side of the debate were his scientific observations—recently made with the first telescopes—of the planets, sun, and moon. On the Vatican's side were centuries of scholarly tradition based on everyday experience and centuries of church tradition based on Scripture. The church condemned Galileo for heresy that day: his writings were banned and he was confined to house arrest.

What led to the conflict between Galileo and the church? In chapters 2 and 3 we found a fundamental harmony and connection between science and Christianity. God created and governs the natural world around us; science is a systematic way to study God's creation. It seems as though there should be no conflict between the two. So what should we do when a Bible verse says the earth holds still and scientific investigation says the earth moves?

In this chapter we'll discuss the source of such conflicts, using the metaphor of two "books" of revelation: God's Word (Scripture) and God's world (nature). We'll argue that

▶ the conflict is not between God's two revelations, but at the human level, between science and the interpretation of Scripture.

▶ science and biblical interpretation do not operate in a vacuum but are influenced by the culture around them.

▶ both science and biblical interpretation can be reliable when good methods are used.

We'll conclude with a description of Galileo's seventeenth-century conflict that illustrates many of the dynamics involved in twenty-first-century conflicts.

THE TWO BOOKS OF REVELATION

In chapter 1 we wrote about God's Word and God's world. God has given us his written Word in the pages of the Bible (the book of Scripture). God has also created the earth and the entire universe (the "book" of nature). Scripture is called God's special revelation, while nature is part of God's general revelation. The Belgic Confession, one of the creeds written at the time of the Reformation, puts this distinction eloquently:

> We know God by two means: First, by the creation, preservation, and government of the universe, since that universe is before our eyes like a beautiful book in which all creatures, great and small, are as letters to make us ponder the invisible things of God: his eternal power and his divinity, as the apostle Paul says in Romans 1:20. All these things are enough to convict men and to leave them without excuse. Second, he makes himself known to us more openly by his holy and divine Word, as much as we need in this life, for his glory and for the salvation of his own.
> —Belgic Confession, Article 2.

This analogy of two books of revelation goes back centuries before the Belgic Confession to Augustine and other theologians. St. Augustine advised Christians to "read" both books:

> It is the divine page that you must listen to; it is the book of the universe that you must observe. The pages of Scripture can only be read by those who know how to read and write, while everyone, even the illiterate, can read the book of the universe.
>
> —Enarrationes in Psalmos, 45.7.

These early theologians drew this idea from Bible passages that attest to God's hand at work in the natural world. Psalm 19 is a beautiful example of this. The first half praises God for his revelation in nature ("The heavens declare the glory of God"); the second half for his revelation in Scripture ("The law of the Lord is perfect, reviving the soul").

The diagram on page 74 illustrates these two books of revelation. The top part of the diagram reminds us that God is the author of both nature and Scripture. *Because God is the author of both revelations, we believe that nature and Scripture do not conflict with each other.*

Revelation is the manner in which God makes himself known to us. Therefore, the "two books" of revelation are, first and foremost, two ways of learning *about God*. God does not reveal contradictory things *about himself* in nature and in Scripture. God is not false or changeable, and we do not expect God to contradict himself by revealing something in nature that is contrary to Scripture. Based on what we learn about God from these revelations, it's reasonable to extend this analogy to knowledge *about creation*. If Scripture has anything to say about the natural world, then what is revealed in Scripture should not contradict what is revealed by studying the natural world itself.

While the "two books" analogy is not perfect, it is useful for understanding conflicts between science and religion. Science is our human attempt to understand the natural world. Biblical interpretation is our human attempt to understand the Bible.

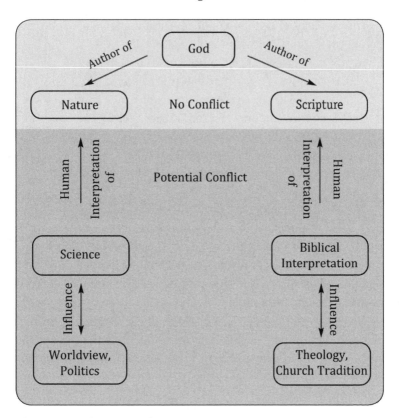

Conflicts can arise because our human understanding of one or both books may be in error.

The bottom boxes of the diagram remind us that science and biblical interpretation do not happen in separate compartments from the rest of our lives. The conclusions of science are influenced by the larger context of our human experiences, including worldviews and politics; the conclusions of biblical interpretation are influenced by theology, church tradition, and even politics, as we'll see later in this chapter as we look at Galileo's story.

CULTURAL INFLUENCES ON SCIENCE

Worldviews and Science Influence Each Other

Worldviews can influence science. As we discussed in chapter 2, a scientist's work relies on worldview beliefs, such as the regularity of nature and the need for experimental testing. Science, in turn, can influence worldviews in healthy ways by providing information and context for making decisions. For example, it's a person's *worldview* that prompts her to care about the environment, but it's *science* that informs her choices about what actions are needed to best care for the environment.

Worldview beliefs can influence the scientific process in another way. When a scientist is choosing between two or more competing models (theories) and the experimental evidence is inconclusive about which is best, beliefs and knowledge outside of science can sometimes influence the scientist's choice. Because the data alone do not favor one model over another, scientists with different worldviews can look at the same data and disagree about which model is best.

Worldviews and science can also interact in less healthy ways. One unhealthy interaction happens when someone rejects a scientific conclusion without examining the data carefully because that conclusion seems to conflict with his or her worldview. Alternately, someone might believe a model not because scientific data actually support it but because it *matches* his or her worldview beliefs. For instance, some practitioners passionately believe that certain kinds of alternative medicine therapies are effective in spite of scientific evidence to the contrary. They want the therapies to work because of their worldview beliefs, in some cases even claiming that their therapies are scientific when the scientific data are against them.

This is where the self-correcting features of the scientific process can help: scientists of differing worldviews challenge each other, forcing each side to make a stronger *scientific* case for its models and inspiring each other toward creative thinking. They invent new technologies and new experiments to support or dis-prove competing models until they reach a new consensus. The competing models and original arguments may have begun, at least in part, because of worldview beliefs, but eventually the experiments and observations push the scientific community toward a consensus shared by scientists of many different world-views.

A number of Christians today accuse the scientific commu-nity of having an atheistic bias on issues of origins *without* first carefully examining the evidence that has led the scientific com-munity to its conclusion. This is an invalid accusation for several reasons:

▶ First, many scientists are not atheists. When the scientific community really does have a consensus, it represents the professional judgment of people with many different religious views, including many Christians.

▶ Second, recall from chapter 2 the idea that all truth is God's truth. Regardless of the worldview beliefs of the person who discovered the scientific truth, if it is true, that knowledge is a gift from God.

▶ Third, we should not be quick to deny a scientific result simply because it disagrees with what we already believe. An apparent conflict should certainly prompt us to demand a solid explanation of the scientific evidence. But a quick rejection does not give sufficient respect to God's revelation in nature since it denies that new truths may be learned from it.

Another unhealthy interaction occurs when science is misused to argue for a particular worldview. For example, atheists—both scientists and nonscientists—have a long history of loudly claim-ing that the results of science prove that atheism is true. When atheists make such claims in their writing and speaking, they are seldom careful to differentiate where the science ends and their worldview claims begin. They tend to thoroughly mix scientific results with their worldview claims so that it is difficult for a non-scientist to tell the difference. This type of writing and speaking has caused the entire scientific community to acquire an atheistic reputation, even though only a few scientists mix atheism with science in this way.

Let's look at an example of this sort of science and worldview mixing, one that might be used in reference to Galileo. An atheist might argue something like the following:

▶ Premise 1: Christianity requires that the earth is stationary in space.

▶ Premise 2: Science proves that the earth moves.

▶ Conclusion: Christianity is false.

Christians clearly disagree with the conclusion, but what is the best way to make a counterargument? Some argue against the conclusion by claiming that the *science* itself is faulty or biased. In the case at hand they would argue that the astronomical evidence is wrong and that the earth really is stationary in space as the Bible says. But it is a grave error to assume automatically that science itself must be the problem. Rejecting the science without first examining it is actually a *theological* mistake as well as a scientific mistake. It too easily sets aside the evidence from God's revelation in nature. We need the information from *both* revelations in order to find the truth in such conflicts.

If the science of premise 2 is well supported, a better approach for Christians is to challenge the worldview claim of premise 1. Maybe Christianity doesn't stand or fall depending on what we believe about the motion of the earth. Indeed, it is vitally important that Christians challenge premise 1. If Christians don't challenge this worldview premise, then we are actually agreeing with the atheist that science alone can decide between Christianity and atheism.

The correct strategy in this case is not to attack the science but instead to disconnect the worldview claim from the science. We need to evaluate scientific statements and worldview statements separately, each on its own merits.

People with different worldviews can have significantly different reactions to the same scientific result. For an example, check out "Five Different Worldview Interpretations of One Scientific Result" on our website (www.faithaliveresources.org/origins).

Politics and Science Influence Each Other

Discussions about science and religion that turn political get even more tangled. People advocate views on the basis of party affiliation and political maneuvering rather than on scientific evidence. Unfortunately, the arguments are still cloaked in scientific language on all sides, and scientific results are often twisted and distorted to make political points. Issues that raise only mild conflicts in science and religion provoke major political battles among individuals or cultural wars between political parties. In the realm of politics and the courts, citizens are often presented with only two choices: siding with the plaintiff or with the defendant, with one political candidate or the other, with one party's platform or the other. Here especially we must remember that usually more than two options exist.

Often if we do some research we can find more common ground and creative answers. For example, a common political fight these days is between those who want to build new power plants or business parks or housing developments and those who want to prevent construction in order to preserve the environment. It often sounds as though we have to choose one of two options: promote economic development while destroying the environment or protect the environment while stifling economic development. But creative people who study the situation can often find options that allow careful construction in ways that preserve the environment.

When science and religion debates turn into political fights, pay special attention. Look for any unquestioned underlying assumptions (such as premise 1 in the earlier example). Ask yourself these questions:

▶ Do the two sides actually care about the scientific data?

▶ Do they care what the religious teachings truly are?

▶ Or has the argument degenerated to become merely a battleground for some other personal or political fight?

CULTURAL INFLUENCES ON BIBLICAL INTERPRETATION

Now let's turn our attention to Scripture, diagrammed on the right-hand side of the "two books" chart on page 74. As Christians, our belief in the truth and authority of the Bible is fundamental to our faith: "All Scripture is God-breathed and is useful for teaching, rebuking, correcting and training in righteousness, so that all God's people may be thoroughly equipped for every good work" (2 Tim. 3:16-17).

Even though Christians generally agree on the inspiration and authority of the Bible and on the central teachings of Scripture, they often disagree on the exact interpretation of some particular passages and on how those passages relate to the issues we face today. That's because, like science, biblical interpretation does not operate in a vacuum but is influenced by the society around it, particularly by theology and church tradition.

Theology and Church Tradition Influence Biblical Interpretation

Theology and church tradition certainly influence the interpretation of a Bible passage, and vice versa. This influence is largely positive. Theology and church tradition are built on what the Bible says, and they in turn help us understand the Bible. Theology is built upon the whole of the Bible, not just bits and pieces of it. Thus theology reminds us that the interpretation of individual verses should be consistent with teachings on the same topic elsewhere in the Bible. Theology and church tradition help us understand which doctrines are essential to Christian faith and which are less central issues where Christians may disagree. Knowledge of church tradition handed down through the centuries keeps us from "reinventing the wheel" on each passage and helps us avoid pitfalls that have been addressed before. Through church tradition each new generation draws on the wisdom of the past to understand what the Bible teaches.

While church tradition is a great blessing for helping us interpret Scripture, sometimes tradition gets it wrong. When that happens, it can be very hard for the church to adopt a better interpretation. Both the Protestant Reformation and the abolition movement against slavery, to name just two historical examples, show how difficult it can be to convince a large segment of the church that it is interpreting certain Bible passages incorrectly. A challenge to the interpretation of a particular passage can make people feel as though the entire theological tradition is being challenged; and when politics gets thrown into the mix, as it did on those two occasions, people have a vested interest in maintaining their particular interpretation.

This is an area in which different branches of the church can help each other. Just as scientists can spot each other's biases and blind spots, so churches in different cultures can hold each other accountable. The civil rights movement in North America provides a striking example. In the early and mid-twentieth century many white Christians were blind to the injustices being inflicted on black Americans. The Southern Christian Leadership Conference, the inspiring leadership of Rev. Martin Luther King Jr., and the courageous actions of many members of the movement helped many Christians awaken their consciences and improve their understanding of Scripture.

CAN HUMAN INTERPRETATION BE RELIABLE?

Is human interpretation of either nature or Scripture reliable? Let's take another look at the diagram on page 74. It not only highlights God's authority on both sides but also points out a *strategy* for handling conflicts. As Christians, we should not simply assume that science is always correct and therefore ignore the Bible, nor should we assume that our interpretation of the Bible is always correct and therefore ignore the natural world that God has made. When a conflict arises, our strategy should be to examine *both* the scientific and the biblical interpretations more carefully.

This isn't easy. It requires that we learn some of the methods and strategies appropriate to science, and it requires that we learn how to engage in careful biblical interpretation. Let's take a brief look at the issues of scientific and biblical interpretation.

Reliability of Scientific Interpretation

To get a handle on the reliability of scientific knowledge, it's helpful to think about science in terms of its models (theories). Some models are merely educated guesses; others are well tested and rock solid. When scientists debate two different models for the same physical system, it is clear that both can't be right. Often a stage occurs in scientific investigations when experiments and observations are inconclusive and don't clearly distinguish between the models. At this stage scientists may be tempted to claim that an experiment *proves* their model and to ignore data that might *disprove* it. Such behavior leads to errors in their conclusions.

In common speech, a person will say "It's just a *theory*" to mean that an idea is only a guess without any support. Scientists use the word "theory" to refer to models from the early educated-guess stage all the way through to the highly tested and well-established stage. Sometimes a theory is even more reliable than a law. For example, Einstein's "Theory of General Relativity" is a more accurate description of nature than Newton's "Law of Gravity."

But the scientific community has developed methods to reduce such errors. Before any result is published in a scientific journal, the work is evaluated by peer review. In peer review one or more scientists reads the journal article to verify that correct methods were used and documented and that the conclusions are well supported by the data and arguments. If the work isn't up to these standards, it is sent back for revision or rejected altogether. A similar process is used before government agencies or private foundations grant funds to a scientist. After the results are published, others in the scientific community who are interested in the same question often repeat the experiment or calculation for themselves and are quick to inform the community if they get different results. By working as a community scientists catch each other's errors. Thus, when the scientific community has reached consensus, the public can feel fairly confident that the scientists have it right.

The model becomes even more solid as it is used over time. Future experiments continue to use the results of the earlier work in a variety of ways, so a significant problem with the earlier work will become readily apparent. New technologies and methods are invented that test the model in more detail, and independent lines of evidence and reasoning are discovered that support or contradict the model. If a scientific model has community consensus and has been around for some time, we can safely assume that it is quite reliable.

Every once in a while, new discoveries cause a good scientific model that has been around for some time and has the consensus of the scientific community to be replaced by a better model. The old model is not tossed out merely because a new model explains the new discovery—the new model must also do as well as the old model at explaining the whole system.

> For one such example,
> check out how Albert Einstein's new understanding of gravity replaced Newton's model. Click on "Gravity and General Relativity" on our website (www.faithaliveresources.org/origins).

Reliability of Biblical Interpretation

The Bible is not merely a book of moral lessons or a simple list of instructions and factual statements. Its authority comes from God. Scripture connects God's action in real historical events with his purpose behind those actions. Scripture also draws us into the story, so that we are not mere readers but citizens of God's kingdom and thus part of the long arc of God's redemptive history.

The Bible is the result of human-divine partnerships in which God *inspired* and commanded human authors to communicate his Word to his people. This divine partnership with human authors meant that Moses, Paul, and all of the other human authors of Scripture wrote using their own style and reflecting their own personality, knowledge, language, and culture. They used a variety of forms, such as letters, parables, proverbs, and poems, including some styles of literature that we don't commonly use today but that were common two or three thousand years ago. God used all of this to faithfully communicate the story of salvation to each succeeding generation.

But how do we know what a particular passage of Scripture means for us today? How do we get at its true message without reading into it what we want to hear? Following are two principles that lend reliability to biblical interpretation:

▶ **Each passage should be interpreted in light of the rest of the Bible.** It is foolish to cut individual verses out of the Bible and apply them to everyday life simplistically or literally without first determining how they fit into the context of the passage, the book, and the entire Bible. If a particular interpretation of a passage leads to a conclusion that is contrary to other parts of Scripture, that's an indication that the interpretation may be incorrect.

▶ **To best understand what the text means for us today, we need to understand how the original author and the intended audience understood it.** At first this may seem like a daunting task since we speak a different language and live thousands of miles and thousands of years apart from the biblical authors. But we can do so by

  ▶ using a good translation that captures the meaning and nuances of the original language.

  ▶ looking at the internal content of the text. For example, is the verse a metaphor, sarcasm, commandment, minor detail, major theme, or something else?

  ▶ looking at the literary genre of the text. Is it written as history, parable, poetry, prophecy, or something else?

  ▶ considering the cultural and historical context of the passage. How did this culture and the surrounding cultures view the topic? What historical events preceded and followed this writing? Was the writing in response to a particular event? Was it written in response to a particular widespread cultural belief of the time?

Let's look at two examples of these principles in action. Amos 4:4 reads, "Go to Bethel and sin; go to Gilgal and sin yet more." Interpreted literally, this is a command to sin. But the first principle indicates that a literal interpretation is incorrect here because it clearly contradicts the rest of Scripture. The second principle leads us to a better interpretation. The context of the whole chapter shows that the tone of Amos 4:4 is sarcastic. It's part of a longer plea by God for his people to turn from sin or suffer judgment. Thus the message for the original audience also applies to us today: we need to turn from our sins and return to God. In this case application of the two principles results in a nonliteral interpretation.

In contrast, the application of the same two principles to Luke 1:1-3 leads to a literal interpretation. Luke begins his account of Jesus' life with these words:

> Many have undertaken to draw up an account of the things that have been fulfilled among us, just as they were handed down to us by those who from the first were eyewitnesses.... With this in mind, since I myself have carefully investigated everything from the beginning, it seemed good also to me to write an orderly account for you, most excellent Theophilus.

A literal interpretation of this passage does not contradict other parts of the Bible, so it satisfies the first principle. Following the second principle, let's consider how Luke and his first readers understood this text. The internal content of the passage has all the earmarks of a nearly modern style of historical writing, including an explanation of sources and research methods. The cultural context of the first-century Christian church shows that this gospel was understood and acted upon as a literal, historical account of what happened. According to the second principle, we too should understand the events recorded in Luke's book as actual historical events, including the death and resurrection of Jesus Christ, even though science cannot explain the miracle of someone rising from the dead.

These principles of biblical interpretation help us in several ways:

▶ They guide us to non-literal interpretations of some passages and to literal interpretations of others. These principles do not lead down a "slippery slope" to a non-literal interpretation of the whole Bible or to a literal interpretation of every passage.

▶ They discourage us from reading the text as though it were written in the twenty-first century. We should not assume that the words and phrases in a text written thousands of years ago in another language will automatically have the same meaning and implications they would have if they were written by an author today in our culture.

▶ They give us a consistent approach to understanding the Bible. By regularly referring to other parts of the Bible and to the first audience for the text, we have a consistent way to decide whether a passage is literal or figurative, historical or symbolic.

Do these principles of biblical interpretation make it too difficult for the average person to understand the Bible? The Protestant reformers emphasized the *clarity* of the Bible, insisting that its primary message is clear to any Christian who can read a good translation. The Westminster Confession, one of the confessions that arose out of the Reformation (www.reformed. org/documents/wcf_with_proofs/), says it this way:

> All things in Scripture are not alike plain in themselves, nor alike clear unto all: yet those things which are necessary to be known, believed, and observed for salvation are so clearly propounded, and opened in some place of Scripture or other, that not only the learned, but the unlearned, in a due use of the ordinary means, may attain unto a sufficient understanding of them (I.VII).

The principle of clarity applies to the *primary* message of the Bible and what is central to salvation. If we want to understand the *details* of how particular Bible passages relate to modern science, we need to do our homework, following the principles discussed above.

THE TWO BOOKS OF REVELATION IN GALILEO'S DAY

To conclude this chapter we'll use Galileo's experiences as a case study for understanding the so-called "war" between science and religion. What events led to his trial in 1633? What issues were at stake?

Galileo's story is much more complex than the simple caricature given at the beginning of this chapter. The conflict was about more than science and the Bible; it also involved academic squabbles, church tradition, personal grudges, and politics. We'll see that Galileo's seventeenth-century story illustrates many of the dynamics involved in twenty-first-century conflicts between science and religion.

Science: The Earth Moves through Space

Before the time of Copernicus and Galileo, the accepted model of the solar system was *geocentric*. In this model the Earth was stationary while the planets, the Sun, and the Moon orbited Earth. This model was developed in great detail by the Greek scientist Ptolemy around A.D. 200, long before the telescope had been invented. Ptolemy worked out the size and speed of each planet's orbit such that it matched the best eyeball observations of how the planets move through the constellations across the sky. Ptolemy's model gave accurate predictions for well over a thousand years for where each planet would be in the sky on future dates.

In the Middle Ages scholars began to notice that the planets were not exactly where the geocentric model predicted they would be. Copernicus (1473-1543) set out to develop a new model for the motion of planets. He could have simply refined Ptolemy's geocentric model by tweaking the planetary orbits a bit; instead he developed a completely new *heliocentric* model. According to his new system the Sun is stationary, the Earth and all the other planets orbit the Sun, and the Moon orbits the Earth. This was a revolutionary idea!

Until Galileo's day most people thought that the earth didn't move. After all, it doesn't *feel* like it's moving, except during earthquakes. But as you read this book, you really are moving rapidly through space because of the many motions of the earth. For more on these motions, visit our website (www.faithaliveresources.org/origins) and click on "Motions of the Earth Through Space."

Thus, in Galileo's day scholars had at least two different models—geocentric or heliocentric—for the motion of the planets. It wasn't at all clear which model was better. As often happens in science, new technology made a big difference. In 1609 Galileo learned about primitive telescopes. After making several modifications to improve their optics and usefulness, he pointed one at the heavens. What he saw was astounding. The planets were not merely points of light moving past the stars but whole worlds themselves. Saturn had rings. Jupiter had stripes. The Milky Way was not merely a foggy band of light in the sky but countless individual stars. Interestingly, Jupiter had four bright moons that orbited around it. These moons clearly did *not* orbit the Earth, a blow against the geocentric model.

The motion of the Earth in the heliocentric model should have caused an optical illusion called *parallax,* in which nearby stars appear to move slightly as the Earth orbits the Sun. This effect was too small to detect with the telescopes of Galileo's time. Opponents of the heliocentric model used this to argue that the Earth was not actually moving. To learn more, visit our website (www.faithaliveresources. org/origins) and click on "Parallax and Its Role in the Heliocentric/Geocentric Debate."

Galileo's telescopic observations of Venus turned out to be critical. What Galileo observed was that Venus moved between full and crescent phases in a cycle that matched the prediction of the heliocentric model. His observations clearly contradicted the geocentric model that predicted only new and crescent phases. Here was firm scientific evidence that the geocentric model was in serious need of repair or was wrong altogether. Galileo became convinced that the Earth moved around a stationary Sun and that his observations proved it.

> For more about the phases of Venus and illustrations, click on "The Phases of Venus" at our website (www.faithaliveresources.org/origins).

Biblical Interpretation: The Earth Is Firmly Established

Let's return to biblical interpretation as shown on the right side of the two-books diagram (p. 74) and consider this aspect of Galileo's situation. Various Bible passages, including Joshua 10:12-14 and Psalm 19:4-6, speak of the motion of the sun or the stability of the earth. These verses seem to support the geocentric model and contradict the heliocentric model. Consider also a phrase that appears in several places: "The world is firmly established; it cannot be moved" (1 Chron. 16:30; Ps. 93:1; Ps. 96:10).

Before Galileo's day Christians read these verses literally—as signifying that the earth *did not* move. But now we have good scientific evidence that the earth *does* move. Most Christians today no longer think these verses should be interpreted literally. Some Christians simply ignore them, figuring that this issue isn't important and that science must have the right answer. But if we ignore every verse in the Bible that appears to disagree with current scientific understanding, we might end up tossing out many important passages, such as those proclaiming miracles and the resurrection. Moreover, this practice would force us to change our interpretation of the Bible every time science changes or develops. Other Christians say that this verse was not intended to be understood literally because it is part of a poem. But poems are not always metaphorical (consider Israel's history recorded in Ps. 78), nor is all prose literal (consider Jesus' parables). While the genre of the text (poem or prose) can be crucial to a good interpretation, it does not *automatically* determine whether or not the text is intended to be understood literally. Still other Christians read the phrase "firmly established" to mean that the Earth is firmly established in its orbit around the Sun. This is appealing for modern readers in that it does match modern science. As we discussed in chapter 2, God sustains the law of gravity and establishes the planets in their orbits.

The problem with all of these interpretations is that they interpret the verses with a modern mindset and immediately compare it to what modern science says. Rather, we should use the principles of interpretation, starting with what the passage meant for the first audience. Certainly the ancient Hebrews had no idea that the Earth orbited the Sun, and it is unlikely that God meant to teach them about planetary orbits. (They didn't even know the earth was spherical!) For them, saying that "the world is firmly established" (Ps. 93:1) would have been a simple observation of the everyday experience of the solid earth under their feet.

Consider this verse in the context of Psalm 93:

The LORD reigns, he is robed in majesty;
    the LORD is robed in majesty and armed with strength;
indeed, the world is established, firm and secure.
    it cannot be moved.
Your throne was established long ago;
    you are from all eternity.

The seas have lifted up, O LORD,
    the seas have lifted up their voice;
    the seas have lifted up their pounding waves.
Mightier than the thunder of the great waters,
    mightier than the breakers of the sea—
    the LORD on high is mighty.

Your statutes, LORD, stand firm;
holiness adorns your house for endless days.

This psalm as a whole is about God's authority. Notice that not only the *world* is firmly established; so are God's throne and his statutes. (Ps. 96:10 and 1 Chron. 16:30 also place this line in the context of describing God's reign.) In Psalm 93 the verse is part of a two-verse couplet, a common device in Hebrew poetry for making two related statements. Here the parallel statements closely identify the world with God's throne. Just as no human action can move the earth under our feet, so nothing can dislodge God from his throne. The stability of the world is used as a fitting symbol for the stability of God's reign. This symbol appears frequently in the Old Testament, in verses like Isaiah 66:1, "Heaven is my throne, and the earth is my footstool."

When the original audience heard these verses, they would not have heard the author trying to *teach* them that the earth was fixed in place. They already believed that! They would have heard the author referring to this common belief as a vivid portrayal of the theological truths of God's authority and the stability of God's reign. As evangelical biblical scholar Peter Enns writes,

> When God reveals himself, he always does so to people, which means that he must speak and act in ways that they will understand. . . . That the Bible, at every turn, shows how "connected" it is to its own world is a necessary consequence of God incarnating himself.
> —*Inspiration and Incarnation,* p. 20.

In other words, God spoke to people using common concepts that they could understand so as not to cloud the intended spiritual message. Galileo felt similarly about this and other Bible passages that alluded to astronomy. Quoting Cardinal Baronius, he quipped, "The Bible teaches how to go to heaven, not how the heavens go."

For more of Galileo's views on biblical interpretation, visit our website (www.faithaliveresources.org/origins) and click on "Excerpts from Galileo's *Letter to the Grand Duchess Christina: Concerning the Use of Biblical Quotations in Matters of Science.*"

This apparent conflict between science and biblical interpretation seems easily resolved. So why did the conflict become inflamed instead?

Worldview, Politics, and Church Tradition in Galileo's Time

Worldview and philosophy played a major role in this conflict. Galileo's earlier studies of motion, buoyancy, and friction contradicted centuries of belief based on Aristotle's teachings about the nature of the universe. This led to a vigorous academic debate at his university, which then spilled over into contentious university politics. This academic debate and his later fights in the church turned nasty in part because of Galileo's arrogant and short-tempered personality.

The tide of church opinion was actually supportive of Galileo in the first few years after he presented his argument for heliocentrism. The church became more defensive when Galileo argued for a different interpretation of Scripture. Centuries of church teaching had been tied to Aristotle's physics and the idea of a stationary earth, and church leaders were not ready to alter those teachings quickly. Moreover, the Roman Catholic Church was on the defensive politically after the Protestant Reformation, particularly on issues of church authority. The church was not so much anti-science as pro-tradition, opposing the Reformation idea that ordinary people could interpret the Bible for themselves.

By 1616 the conflict had heated up, and the Holy Office declared that the concepts of a stationary sun and a moving earth were heretical. Galileo was instructed to desist from advocating the heliocentric model, although he could study it hypothetically.

He agreed to this, but in the 1620s he began writing his *Dialogue on the Two Principal World Systems*. The premise of this book is a debate in which characters argue about the geocentric and heliocentric models. Since the argument for heliocentrism was couched in the words of a fictional character and opposing arguments were included in the words of other characters, Galileo could claim he was not advocating the view himself.

In 1631 the church gave official approval for the book to be published, since the debate was hypothetical. But when the book was widely read it became obvious that Galileo hadn't followed the spirit of the injunction in 1616. The arguments for heliocentrism were strong, while the arguments for geocentrism were weak. Moreover, Galileo had antagonized the Pope by putting one of the Pope's own arguments for geocentrism in the mouth of a character named "Simplicio," in effect calling the Pope "stupid." In 1633 Galileo was brought to trial for advocating heresy.

Thus the underlying conflict about geocentrism and heliocentrism, pitting as it did scientific models against biblical interpretation, was fanned into flame by worldview conflicts, academic politics, church tradition, and personal animosity. Although the underlying conflict is readily resolved today, the harsh debates of the seventeenth century have left the church with an undeserved anti-science reputation that has lasted into the twenty-first century.

In 1642 Galileo died under house arrest, at the same time that interest in his discoveries was building around Europe. That same year saw the birth of Isaac Newton, who went on to publish his *Universal Law of Gravity* in 1687. With Newton's understanding of gravity it makes sense that the planets do in fact orbit the Sun— the most massive object in the solar system. Newton's work, along with improvements in telescope technology and more accurate astronomical observations, brought about a scientific consensus in support of heliocentrism. The parallax of stars was finally observed in 1838, confirming the motion of the Earth.

The Vatican officially withdrew its condemnation of Galileo in 1992, after having unofficially changed its view much earlier. Most publications promoting the Copernican view were dropped from the list of banned books in 1757, and in 1822 Galileo's *Dialogue* was dropped from the list. In 1891 the Roman Catholic Church founded the Vatican Observatory, which is still in operation today. The Church continues to fund cutting-edge astronomical research, a testimony to its commitment to science as the exploration of God's creation. In 1991 Pope John Paul II wrote, "Science can purify religion from error and superstition; religion can purify science from idolatry and false absolutes. Each can draw the other into a wider world, a world in which both can flourish" (*On Science and Religion*).

This positive interaction between science and religion is sometimes lost in the midst of apparent conflicts. As in Galileo's time, conflicts can become hurtful on an interpersonal level, as well as difficult to resolve intellectually. The ideas portrayed in the "two books" diagram (p. 74) are helpful in both areas. Christian unity is strengthened when Christians on all sides agree that proper understandings of both nature and Scripture are essential to the solution, as well as that the conflict is at the human level rather than between nature and the Bible itself. We can persevere, knowing that a resolution of the conflict must exist even though we haven't found it yet. Because God does not contradict himself, nature and Scripture cannot be in actual conflict. So, rather than ignoring one or the other, we study both with the hope that ultimately we might grasp the underlying truths that God has revealed.

QUESTIONS FOR REFLECTION AND DISCUSSION

1. In what ways are nature and Scripture alike? In what ways are they different?
2. In what ways are science and biblical interpretation alike? In what ways are they different?
3. If you're not a scientific expert, how can you tell how reliable a scientific result is? What should you look for?

4. What do you think of the principle "To best understand what the text means for us today, we should first seek to understand how the original author and the intended audience understood the text"? Does some Scripture have hidden content or meaning for us today that would have been incomprehensible to the original audience? Might Scripture today ever mean something *different* from what it meant to the original audience?

5. In Mark 4:31 Jesus claims that the mustard seed is the smallest seed on the earth, but other types of seeds have since been found that are smaller. Is this a conflict between nature and Scripture? If so, how could this conflict be resolved?

6. Some people today refer to Galileo's work as the first in a long line of scientific discoveries in which science has triumphed over religion and superstition. How would you respond?

7. When would it be appropriate to let science affect biblical interpretation? When would it be inappropriate?

8. When would it be appropriate to let biblical interpretation affect our understanding of the natural world? When would it be inappropriate?

9. What similarities do you see between the Galileo incident and modern debates about creation, evolution, and design? What differences?

ADDITIONAL RESOURCES

More on biblical interpretation:

Berkhof, Louis. *Principles of Biblical Interpretation.* Grand Rapids, Mich.: Baker Book House, 1950, 2003.

Fee, Gordon, and Douglas Stewart. *How to Read the Bible for All It's Worth,* 3rd edition. Grand Rapids, Mich.: Zondervan, 2003.

More on the relationship between Christianity and science:
Barbour, Ian G. *Religion and Science.* New York: Harper, 1997.

Enns, Peter. *Inspiration and Incarnation: Evangelicals and the Problem of the Old Testament.* Grand Rapids: Baker Academic, 2005.

Lindberg, David C., and Ronald L. Numbers. "Beyond War and Peace: A Reappraisal of the Encounter between Christianity and Science," *Perspectives on Science and Christian Faith,* 39.3:140-149 (1987).

More on the Galileo story:
Drake, Stillman, ed. *The Discoveries and Opinions of Galileo.* New York: Doubleday, 1957. Includes translation of Galileo's "Letter to the Grand Duchess Christina."

Gingerich, Owen. *The Book Nobody Read: Chasing the Revolutions of Nicolaus Copernicus.* New York: Walker & Company, 2004.

Hummel, Charles E. *The Galileo Connection.* Downers Grove, Ill.: InterVarsity Press, 1986.

Numbers, Ronald, ed. *Galileo Goes to Jail and Other Myths About Science and Religion.* Cambridge, Mass.: Harvard University Press, 2009.

CHAPTER 5

GENESIS: CONCORDIST INTERPRETATIONS

O ne May, while living in Massachusetts, we enjoyed walking along the rugged Atlantic sea coast at Gloucester and Rockport. As we watched gulls sweeping low, waves broke over the large boulders and shelves of rock. Looking closer, the history embedded in those rocks seemed to jump out at us.

We saw regular, thin stripes in shades of gray. These were sedimentary layers, formed from the muddy sediment at the bottom of a lake or sea. We also saw sparkling pink veins of granite, an igneous rock formed when hot magma slowly cools in chambers underground. We noticed that the sparkling granite veins were at odd angles compared to the gray sedimentary layers, cutting across them without destroying the thin layers. Thus, the magma to make the granite must have arrived after the muddy sea bottom had hardened into solid rock. The motions of the earth's crust must have moved the sedimentary seafloor deep under the surface, where it was exposed to the hot magma. And after that, the earth's crust must have moved again to bring the whole mass of rock up to the surface, tilting it so that the interesting layers were exposed. The pounding of the ocean waves then eroded the rock, smoothing away the rough edges. As we looked at the layers and colors of rock, we marveled at how many different processes God used to make them, each process occurring in a particular sequence and each working slowly over a long period of time.

Genesis 1:9-12 states that God created the dry land: "And there was evening, and there was morning—the third day" (v. 13). These words imply that God made all the lands of the earth, including those rocks on the Massachusetts sea coast, in a single day. But the rocks themselves look as though they were formed by processes that would have required millions of years.

Like Galileo's discovery of the earth's motion, these rocks present an apparent discrepancy between the natural world and Scripture. As argued in chapter 4, the apparent discrepancy is not a conflict between God's revelation in the rocks themselves and God's revelation in Genesis 1. Rather, the conflict must be at the level of human *interpretation* of the rocks, the Bible, or both. To resolve this conflict, both sides must be examined—both the biblical interpretation of Genesis and the scientific interpretation of the rocks (geology).

In this chapter we'll focus on *concordist* interpretations. In the concordist view God made the earth using the sequence of events described in Genesis 1. In chapter 6 we'll focus on *non-concordist* interpretations. According to a non-concordist view God created the earth using a different timing and order of events than those described Genesis 1. (The summary of interpretations on p. 100 will give you an overview of both chapters.)

BIBLICAL TEACHING ABOUT THE CREATION OF THE NATURAL WORLD

From New Testament Passages

We'll start by looking at what the Bible teaches about the origin of the world. While most of the attention in this conflict is focused on Genesis 1, it's useful first to look at some New Testament passages about creation. Before reading further, take a few minutes to read John 1:1-3; Colossians 1:15-20; Hebrews 1:1-4; and Hebrews 11:3.

These verses, also summarized in various creeds and confessions of the church, teach several important doctrines about creation:

▶ Christ plays an important role in creation and providence. The forming and sustaining of the world was and is the work of the Trinity, not just the work of God the Father.

▶ God created everything. Every corner of this world, all matter and energy, even space and time, were made by God. You won't find anything in the universe that was made by some other power or authority.

▶ God created from nothing (Latin: *creatia ex nihilo*). Matter and energy are neither self-existing nor coeternal with God. They were made by God. The fabric of space and even time itself does not exist apart from God. God did not use preexisting "ingredients" to create but created all that is without any precursor.

▶ God sustains everything (Latin: *creatia continuans*). God directly upholds the ongoing function of the physical laws, the fabric of space and time, and the continuing existence of matter and energy. Without God's providential hand the laws of nature and matter itself would fall apart. (This is opposed to the worldview of deism, which says that the universe can exist on its own and that God's role was only to get it started.)

Notice that these New Testament passages teach us things about creation, particularly about the role of Christ in creation, that are not found in Genesis 1. Theologians call this *progressive revelation* in Scripture. God did not reveal everything at one time to the first biblical authors and their ancient audiences; he revealed additional information in stages.

From Old Testament Passages

Now take a moment to read three Old Testament passages that teach about the origin of the world: Psalm 104; Genesis 2:4-25; and Genesis 1:1-2:3. (Note that the first three verses of Genesis 2 are the conclusion of the account in Genesis 1. When we refer

Summary of Several Interpretations of Genesis

Following is a brief description of interpretations of Genesis we'll discuss in chapters 5 and 6. The interpretations are not entirely unique but have some overlap. Some of the interpretations, particularly the non-concordist ones, can be combined without contradiction.

Concordist Interpretations

Young Earth Interpretation
Creation occurred about 6,000 years ago, during six 24-hour days, in the order described. A scientific study of the earth should confirm this.

Gap Interpretation
Earth was created long ago (Gen 1:1), became "formless and empty" (Gen 1:2), and was restored about 6,000 years ago during six 24-hour days.

Day-Age Interpretation
Creation occurred over billions of years. Each "day" of Genesis 1 corresponds to a long epoch. Events occurred in the order given in the text, but stretched out over a longer time period.

Appearance of Age Interpretation
Creation occurred about 6,000 years ago during six 24-hour days, but it was created to look like it had a long history of billions of years.

Non-concordist Interpretations

Proclamation Day Interpretation
The days of Genesis 1 took place in God's throne room, wherein God proclaimed each step of creation. The throne-room days are not related to days or time periods on earth.

Creation Poem Interpretation
The number and ordering of the "days" of Genesis 1 are chosen for poetic and thematic reasons rather than historical reasons.

Kingdom-Covenant Interpretations
As the great King, God creates the "realms" of his kingdom; humans are given dominion as in a "land grant" covenant and are delegated divine authority.

Ancient Near Eastern Cosmology Interpretation
Genesis 1 matches the physical picture of the world believed in Ancient Near East religions, but presents a dramatically different theological picture, proclaiming one God as creator of all rather than many gods.

Temple Interpretation
God inaugurates the cosmos as his temple. The six days establish the *function* of each creature and do not refer to the formation of physical material.

to Genesis 1 in this chapter, we're also including the first three verses of Genesis 2.) As you read, think about the similarities and differences between these passages and how they compare to the New Testament passages.

One obvious similarity among these three Old Testament accounts is that they describe the same creation event. They all teach that there is *one* Creator who made everything and that everything created is good and orderly. But there are clear differences in tone and style. (Notice the different names used for God and the use of prose versus poetry.) The three passages emphasize different parts of the story. (Notice how humans are discussed differently in each passage and the different role of water in each.) These three different ways of recounting the same creation event are analogous to the four Gospel accounts of Jesus' life, each with variations in tone and emphasis.

Another difference among the passages is the chronology of events. The language of Psalm 104 is less chronological, so we'll just compare Genesis 1 and Genesis 2. Notice the order in which God created things in each passage. The table below lists the sequences.

Order of Creation in Genesis 1 and 2

Genesis 1:1-2:3	Genesis 2:4-25
Heavens, earth, waters (vv. 1-2)	
Light (v. 3)	
Sky (vv. 6-8)	
Dry land (vv. 9-10)	Dry land, rivers (vv. 5-6)
Plants (vv. 11-12)	Man (v. 7)
Sun, moon, stars (vv. 14-17)	Plants (vv. 8-9)*
Sea creatures and birds (vv. 20-21)	Land animals and birds (v. 19)
Land animals (vv. 24-25)	Woman (vv. 20-22)
Human beings (vv. 26-27)	

*While the TNIV translates Genesis 2:8 as "God *had planted* a garden," the original Hebrew can also be translated as "God *planted* a garden." Several English translations use the latter phrase, which implies that the garden was made after man.

Genesis 2 is missing several elements present in Genesis 1 (ocean and sea creatures, light, sun, and moon). In Genesis 2 birds are created at the same time as animals rather than the day before, as in Genesis 1. In Genesis 2 man is created before the plants, animals, and birds, whereas in Genesis 1 man is created afterward. If each passage is interpreted as a straightforward historical account of how the earth was created, then these two passages contradict each other on the order of events.

Christians over the centuries have chosen several different interpretations. Some say that Genesis 1 refers to global creation and Genesis 2 to a local event. Some posit that Genesis 2 is an extended description of day 6 of Genesis 1. Some suggest that one or both chapters is a nonsequential retelling of events deliberately presented out of order to make a theological point.

Later in this chapter we will discuss a variety of ways in which Christians have interpreted Genesis 1. For now we note that the differences between Genesis 1 and 2 require all Christians to do some interpreting beyond a simple historical reading. The differences between the passages also make it difficult to compare the Bible to scientific evidence, because the Bible itself has some ambiguity in terms of the sequence of events. Most Christians have ignored the ambiguity in the sequence of events in Genesis 1 and 2 and instead have focused on the length of time (age of the earth) involved in creation.

YOUNG-EARTH INTERPRETATION

A common interpretation of Genesis is that God created the earth in six days. Genesis 1 lists creation in one week. Genesis 2 is less clear about the length of time, but it certainly implies that creation took much less time than a human lifetime. In fact, Genesis 2:4 is sometimes translated "On the day when the LORD God made the earth and the heavens," implying that the creation events in Genesis 2 took place in one day.

The Bible also speaks about the length of time *since* the accounts of Genesis 1 and 2. The Old and New Testaments describe historical events in the nation of Israel and list genealogies of their forebears. If these passages are interpreted as being historically precise, with no gaps in the genealogies, they can be used to calculate backward to determine the year in which God made the universe. Archbishop James Ussher of Ireland (1581-1656) studied these passages, did the calculation with impressive precision, and declared the exact date of creation to be Sunday, October 23, 4004 B.C. Even if the genealogies have gaps, the date of creation wouldn't be more than eight or ten thousand years ago.

Up to the 1600s most Christians believed that the world was created a few thousand years ago. They did not have any evidence from the natural world to suggest that a different interpretation of Scripture might be more accurate. (Based on Scripture, most Christians up to the 1600s also believed that the earth did not move through space, and no scientific evidence at the time indicated otherwise.)

But it's worth noting that some Christians in these pre-scientific times held different interpretations of the biblical passages. Some believed that the days referred to much longer time periods based on the verse "With the LORD a day is like a thousand years" (2 Pet. 3:8; see also Ps. 90:4). Justin Martyr (A.D. 155) and Irenaeus (A.D. 189) noted that Adam was told "in the day that you eat of it you shall die" (Gen 2:17, NRSV), yet Adam lived for 930 years (Gen 5:5). Justin and Irenaeus took this to mean that the *day* in Genesis 2 was symbolic of 1,000 years and that Adam was sentenced to die before 1,000 years.

Saint Augustine (A.D. 354-430) believed that God created everything *instantaneously* rather than taking a whole week. Based on internal inconsistencies in the text of Genesis 1, he and others found the six days very difficult to understand. For instance, God is said to have made light on day one but the sources of light (the sun and moon) on day four. The stated purpose of the sun in day four is to mark the passage of days and seasons. What do *day, morning,* and *evening* mean before the creation of the sun? How do the plants made on day three survive without the sun? Given this evidence in the text itself, Augustine argued that the days in Genesis 1, particularly the first three, should not be interpreted as literal days.

Before the development of modern geology these alternative interpretations were rare. Most Christians held a Young-earth Interpretation of Genesis and believed that the earth was created in six twenty-four-hour days just a few thousand years ago.

THE BEGINNING OF GEOLOGY

In the 1600s, the century when James Ussher and Galileo lived, geologists were beginning to catalog rocks in a systematic fashion. They classified them by type and recorded where they were found. Communication was improving throughout Europe, and this helped scholars share their knowledge and ideas. They began to look for scientific explanations for various types of rocks and for larger geological features such as valleys, mountains, and outcroppings. They watched slow, gradual processes, such as erosion and the build-up of sediment, and began to wonder what effects these could have over the many years of earth's history.

Most geologists in the 1600s interpreted Genesis in the same way that Ussher did. They assumed from the outset that the earth was only a few thousand years old and that a catastrophic global flood occurred during the time of Noah. ("Flood geology" saw a resurgence in the 1960s, but the idea goes back to the beginnings of geology.)

In Britain scientists such as John Woodward and Thomas Burnet worked at developing a natural history that fit the young-earth view with known rock types and locations. They and other early geologists made scientific models for geological features that included the effects of a global flood. For instance, when they saw river valleys that were much wider than the rivers that ran through them, they reasoned that the valleys had been created by the receding waters of Noah's flood and that the rivers today are much smaller than the floodwaters. When they saw fossils of sea shells and fish on high mountains, they hypothesized that Noah's flood had lifted the waters and sea creatures to that elevation. When they saw rocks with many layers (called *strata*), they hypothesized that they had formed from layers of sediment laid down by Noah's flood, while rocks that were unstratified (uniform, without layers) were formed in the original creation. They believed that this global flood model could explain why stratified rocks had fossils and unstratified rocks did not: the fish and animals killed by the flood were embedded in the layers of muck that became stratified rock, while unstratified rock predated the flood. On the basis of this model Woodward predicted that the denser sediment and life-forms would be found in the lower layers because they sank more quickly during the flood, while material that sank more slowly would be found in the upper layers.

These scientists took a common interpretation of Genesis—a recent creation and a global flood—and used it as a scientific model to explain what they saw in nature. Their model explained many geological observations at the time and made predictions that could be tested by later geologists.

> To help illustrate the process of rock formation, we've included a fun activity on our website that you can try with your family at home.
> Click on "Making 'Rocks' from Crayons"
> at www.faithaliveresources.org/origins.

Geological Evidence Against a Young Earth and Global Flood Known by 1840

During the 1700s geologists greatly increased their knowledge of the earth. They discovered many more fossils and learned much more about the types of rocks in which fossils are found. They mapped out the stratified and unstratified rocks over entire mountain ranges and countries. Several mining companies were founded in France, Germany, and Italy, and geologists undertook serious study of the underground layers. They tested their models for natural history against many new locations and many types of geological features. As the new data poured in it became apparent that the model of a young earth and a single global flood, as outlined by Burnet, Woodward, and other early geologists, did not fit this larger body of observations.

Several observations made in the 1700s and early 1800s contradicted the predictions of the young earth and global flood model. Geologists discovered

▶ dense materials in all sedimentary layers. Contrary to Woodward's prediction from 1695, sedimentary rocks (and the fossils they contained) were not stratified with denser material in lower layers and less density in upper layers. Differences between the lower and upper layers were present, but these differences could not be explained by density.

▶ conglomerate rocks. Some rocks indicate multiple floods or at least multiple wet periods. Geologists found samples of stratified conglomerate rock in which smooth, rounded pebbles were embedded in the layers of fine-grained sediment. In itself, this is not surprising; a pebble could settle into the layers of mud that later hardened into sedimentary rock. But in some cases the pebbles themselves were found to be a different type of sedimentary rock, such as sandstone or limestone. Thus the pebble itself must have formed in an earlier wet period of sedimentation, dried, and hardened into rock, and the rock broken apart into pebbles. Then, in a later flood or streambed, water eroded the pebble to a smooth surface, and it settled with other sediment to make the conglomerate rock. Rocks like this could not have been formed during a single global flood.

▶ very thick layers of sedimentary rock. Some stratified rock contains so many layers that it is hundreds of feet, or even miles, deep. Geologists in the 1700s were unable to drill for samples miles below the surface, but they could study mountain ranges in which the rock layers had been tipped at an angle to the horizon and exposed above the surface. In some places the layers were documented to be miles thick when measured perpendicular to the layers. For instance, the sedimentary rocks in the central Appalachians of Pennsylvania are at least 40,000 feet thick. A single, year-long flood would not have eroded enough material to deposit layers that thick.

▶ a long history of volcanic activity. Volcanic cones were discovered under grasslands in south central France. Since no human record or legend tells of volcanoes in that area, the last volcanic eruption must have been before human history. Upon close inspection geologists were able to map multiple layers of lava flows, showing that the volcanoes in that area had erupted repeatedly, hardening after each eruption and forming additional structures. Evidence also shows significant water erosion taking place between the various volcanic eruptions. This area tells of a longer and more dynamic history than could be fit into a few thousand years, even with a flood.

For additional geological observations made before 1840, visit our website (www.faithaliveresources.org/origins) and click on "Centuries of Geological Evidence for an Old Earth."

Any small subset of observations can be explained by multiple models. For example, the observation of seashells on mountaintops can be explained by a global flood, by a former sea floor being lifted to become a mountain, or by other models. So a small subset of observations is not enough to prove that one model is better than the others. But as we discussed in chapter 4, a scientific model is judged not by its consistency with a few observations but by its consistency with *all* of the available data. The global flood model could explain a few observations (like the existence of seashell fossils on mountains), but it was *not* consistent with a majority of the data accumulated by geologists.

Because of this kind of evidence, by about 1840 virtually all practicing geologists, including Christian geologists, believed that the earth must be at least millions of years old. Moreover, if a flood had occurred, it must have been local, not global. The data from many locations indicated that the world's stratified rocks and fossils could not have been deposited in a single global flood. While local floods certainly did take place here and there, a longer time period and more gradual processes are required to explain the entire picture. Scientific study indicated that the earth had a long geological history *before* humans arrived on the scene. Thus, just as in Galileo's time, an apparent conflict existed between what nature seemed to be saying and what Scripture seemed to be saying. The evidence from nature did not match the most common interpretation of Genesis at the time.

Note that these geologists were not atheists who set out to disprove the Bible, nor did they necessarily have a "soft" view of Scripture. Many started out with a firm commitment to interpret Genesis as literal, accurate history, and geologists pursued this view for well over a century. If the rocks of the earth had been consistent with a young earth and global flood model, these scientists surely would have noted this. Instead, the earth itself testified otherwise, over and over again.

But Christian geologists did not abandon the Bible. Nor did they abandon their study of nature or deny what it was telling them. They not only continued to investigate the book of nature but also studied the book of Scripture and considered other ways to interpret Genesis 1.

Most of these new interpretations were *concordist* interpretations, views that God made the earth using the *same sequence* of events described in Genesis 1, while ignoring the different sequence in Genesis 2. In some concordist interpretations the events are stretched over longer time periods, but in all such interpretations the order of events is the same: God first made light, then sky and ocean, then dry land and plants, and so on.

GAP INTERPRETATION

The *Gap Interpretation* (or *Ruin-Restitution Interpretation*) became popular by the 1840s. This view focuses on the meaning of two verses in Genesis 1:

▶ In the beginning, God created the heavens and the earth (v. 1).
▶ Now the earth was formless and empty ... (v. 2).

Proponents of this interpretation view the first verse as a complete statement that God made the whole universe millions or billions of years ago; the verse is not an introduction to what follows. They interpret the second verse to mean that the earth *became* formless and empty because of a more recent catastrophe that destroyed life on its surface. (Some suggest that this catastrophe is related to the fall of the angels and Satan.)

The remaining verses of Genesis 1 are viewed as the restitution—*recreation*, not creation—of the earth and life occurring just a few thousand years ago. Thus a long gap of time existed between the creation of the geological earth millions or billions of years ago and the rest of the creation story. This is a *concordist* interpretation; the length of events in Genesis 1 is stretched out but the order remains the same. This stretching provides time for the long historical record of geology to take place. The rest of biblical history starts less than 10,000 years ago and proceeds onward with the development of plants, animals, and humans.

While the Gap Interpretation resolves conflicts with the great age of the earth evidenced by geology, it contradicts other scientific evidence. Scientists have never found geological evidence of this purported ruin and restitution. Moreover, fossil evidence shows that many life-forms, including most species we see today, have been around for much longer than 10,000 years, predating the catastrophe in which they were supposedly destroyed.

DAY-AGE INTERPRETATION

Another concordist interpretation introduced in the late 1700s is the *Day-Age Interpretation*. In this view the days of Genesis 1 actually refer to long periods of time. Each day is stretched to millions or billions of years to provide time for long astronomical, geological, and biological processes to occur (without the discontinuity of the ruin-restoration interpretation). Some argue for this interpretation on the basis of the seventh day of creation. Since Genesis 2:3 does not contain the phrase "and there was evening and there was morning, the seventh day," they argue that the seventh day never ended but continues until today. If the seventh day is longer than twenty-four hours, then the other days could also symbolize longer time periods.

A more common argument for the Day-Age Interpretation is based on the original Hebrew word *yom*, which can be translated as a twenty-four-hour day or as a long, indefinite time period. The latter meaning is less common in the Old Testament but occurs several times (see Josh. 24:7; Isa. 34:8). One problem with this interpretation is the statement that "there was evening and there was morning" at the end of each of the first six days in Genesis 1. This seems to clearly indicate that *yom* refers to twenty-four-hour days, at least for the first six days. Some people who advocate this interpretation say that this phrase refers to the final twenty-four-hour day at the end of each long *yom*.

While the Day-Age Interpretation resolves the conflict with the *length* of time needed to match the scientific data, it is still in conflict with the recorded *sequence* of events. As discussed earlier in this chapter, Genesis 1 and 2 indicate different orders for the creation events. The best current scientific understanding of natural history indicates a sequence that is different from either Genesis 1 or Genesis 2 (see table on p. 111).

Order of Creation in Genesis 1, Genesis 2, and Modern Science

Genesis 1:1-2:3	Genesis 2:4-25	Modern Science*
Heavens, earth, waters (vv. 1-2)		Matter, energy, space, time (13.7 bya)
Light (v. 3)		Stars (13.5 bya)
Sky (vv. 6-8)		Sun (4.6 bya)
Dry land (vv. 9-10)	Dry land, rivers (vv. 5-6)	Moon, earth, dry land (4.6 bya)
Seed-bearing plants (vv. 11-12)		Oceans (4 to 4.4 bya)
Trees bearing fruit (vv. 11-12)	Man (v. 7)	Single-celled life (approx. 3.8 bya)
Sun, moon, stars (vv. 14-17)		Multi-celled life (1 to 2 bya)
Sea creatures (vv. 20-21)	Plants (vv. 8-9)	Various sea creatures including early fish (approx. 520 mya)
		Non-seed-bearing plants on dry land (approx. 450 mya)
	Land animals (v. 19)	Land animals (approx. 380 mya)
		Seed bearing plants (370 mya)
		Insects (approx. 350 mya)
		Dinosaurs (230 mya)
		Mammals (approx. 200 mya)
Birds (vv. 20-21)	Birds (v. 19)	Birds (approx. 150 mya)
Land animals (vv. 24-25)		Fruit-bearing plants (130 mya)
Men and women (vv. 26-27)	Woman (v. 21-22)	Men and women (see ch. 11)

*The abbreviation *bya* means billion years ago; *mya* means million years ago.

This scientific understanding is a consensus that scientists have held for over a century. The order of events in Genesis simply doesn't fit the order of events reconstructed from a scientific study of nature, no matter how much the time involved is stretched out. In particular, consider what happens on day three.

How could plants survive for millions of years without sunlight for energy, insects to pollinate the flowers, birds and animals to spread their seed, and worms to aerate the soil? The modern understanding of ecosystems indicates that all life-forms in a system (plants, insects, birds, animals) rely on each other, and most modern species cannot survive on their own without the others. In order for this to work, advocates of the Day-Age Interpretation often interpret the list of plants on day three to mean only the very earliest, single-celled forms of life on earth, and the creation of the sun on day four to indicate its first appearance through the cloud layers rather than its initial formation.

APPEARANCE OF AGE INTERPRETATION

Because the Gap Interpretation and the Day-Age Interpretation have difficulty matching *all* of the scientific data, *Appearance of Age*, another concordist interpretation, has been proposed. This interpretation suggests that God really did create the earth in six twenty-four-hour days less than 10,000 years ago but that he constructed it to *look* ancient. Appearance of Age was suggested as early as the 1800s and is still advocated by some Christians today. It neatly avoids all conflict between what nature tells us and what Scripture says about natural history. Unlike other concordist interpretations, the scientific evidence cannot disprove this view: whatever science finds is the way God made it appear. Thus no scientific problems arise with this view.

But there is a significant *theological* problem. This interpretation would mean that God embedded within the universe a host of evidence that indicates a long, richly detailed history that never happened. Sedimentary layers look as though they formed in a muddy sea bottom, except that the sea never existed. Igneous rock layers look as though they were formed from repeated volcanic outflows, but the volcano never erupted. Millions of plants and animals are found in the fossil record, but those organisms never actually lived. Would God have created the earth with evidence of a rich, complicated history that is completely false? We might as well ask "Did God create the world last Tuesday?"

God *could* have created the earth last week, complete with history books on library shelves, decayed statues in museums, and false memories in our brains, but this seems dishonest. The elaborate *detail* of the false history makes the Appearance of Age Interpretation theologically implausible. Of course, God is omnipotent and could have created it this way if he chose, but the Appearance of Age Interpretation seems inconsistent with what we know about God as taught in the rest of the Bible.

"The heavens declare the glory of God" (Ps. 19:1). This verse and many other Scripture passages teach that God reveals himself to us truthfully through the natural world. God doesn't ask us to believe that the earth stands still in spite of abundant astronomical evidence that it orbits the sun. Would God really ask us to believe that the earth is 10,000 years old in spite of abundant geological evidence that it is billions of years old?

AGE OF THE EARTH AND FUNDAMENTALISM IN THE EARLY 1900S

By 1840 Christians were considering a number of different interpretations of Genesis. Each interpretation upheld the divine inspiration and authority of Scripture. Each attempted to find harmony between the revelations of the book of nature and the book of Scripture. The Gap and Day-Age Interpretations were popular among conservative Christians in the 1800s, and the Day-Age Interpretation is still popular today.

In the following decades two major developments affected how Christians thought about the age of the earth and geology. In 1859 Charles Darwin published *On the Origin of Species*, introducing the idea of biological evolution by means of natural selection. (We'll discuss this in detail in chs. 8 and 9.) And in the late 1800s some "liberal Christianity" views—including rejection of the authority of the Bible, freedom to construct nontraditional views of God, and questioning of biblical miracles—became popular.

By 1900 many conservative Protestant Christians and some Catholic Christians had thoroughly rejected liberal Christianity for theological reasons. Some, like Benjamin Warfield, accepted Darwin's theory of evolution as a scientific theory, while others rejected evolution—not for scientific reasons but because they feared that it had unacceptable religious consequences.

In 1915 the views of some leading Protestants were collected in *The Fundamentals*, four volumes of essays by pastors and theologians that greatly influenced the development of the evangelical and fundamentalist movements. (The term *fundamentalist* derives from this book.) The goal of these essays was to lay out "essential" Christian doctrines and to refute liberal Christianity. They roundly condemned "higher criticism" of Scripture, affirming the inerrancy of the Bible and the historical death and resurrection of Christ. These volumes also included several essays addressing contemporary science and several that denounced evolution, particularly the evolution of humans. But even these essays contain little mention of the age of the earth and geology. In fact, issues of age are conspicuous by their absence. If these leaders of the faith had considered a young earth and six-day creation to be essential to Christian doctrine, they surely would have taught it in this volume.

Instead, James Orr wrote of the worth of pursuing astronomy and geology and the consistency of scientific results with Christian faith at a time when the overwhelming consensus of astronomers and geologists was that the universe and the earth are very old ("The Early Narratives of Genesis," *The Fundamentals,* Volume 1). Dyson Hague argued that the historicity of Genesis is essential to Christian doctrine and that "man was created, not evolved." Yet when discussing the first four days of creation Hague wrote, "Genesis is admittedly not a scientific history. It is a narrative for mankind to show that this world was made by God for the habitation of man, and was gradually being fitted for God's children" ("The Doctrinal Value of the First Chapters of Genesis," *The Fundamentals,* Volume 2).

Some Christians today claim that young-earth creationism has been the primary Christian view throughout all of church history, but in fact this view was not widely held by conservative Christian leaders in North America, including those in the fundamentalist movement, from the early 1800s to the mid 1900s.

NEW EVIDENCE FOR OLD AGE

Continental Drift

In the 1900s scientists found several more lines of evidence that the earth is old. One of these is the discovery of *continental drift*. Continental plates move about an inch or two per year; the exact speed and direction of motion is different for each plate. The model of continental drift was proposed in the early 1900s but not immediately accepted by most geologists, especially since no technology was available at that time to measure that motion. By the late 1960s it was clear that this model could explain a host of geological data:

▶ locations of earthquakes
▶ formation of mountain ranges and volcanoes
▶ mid-Atlantic undersea ridge
▶ uplift of former sea beds to great heights

Within the last few decades new technology has made it possible to confirm *directly* the motion of continental plates. Radio astronomers noticed that their radio telescopes on different continents were moving relative to each other. Today global positioning satellites are used routinely to chart the motion of continental plates.

The model of continental drift also allowed scientists to explain the shapes of some continental plates and the locations of certain plant and animal fossils. By projecting the motion of continental plates back in time, scientists calculate that about 180 million years ago the plates fit together to form a single large continent called *Pangaea*. (Imagine puzzle pieces in the shapes of the continents, and notice how Africa fits with North and South America.) Since the time of Pangaea, the plates split apart along the mid-Atlantic ridge to form the Atlantic Ocean. This Pangaea model predicts that the west coast of Africa and the east coast of South America will have similar rock layers with similar plant and animal fossils in the regions that were connected before the breakup. That prediction has been confirmed by numerous geological studies on the two continents. The continental drift model explains a large set of geological observations, and it also indicates that the continents are at least several hundred million years old.

Layering

Another independent line of evidence for an old earth is ice layering in glaciers. Layers are produced on glaciers each year as snow falls and atmospheric dust settles. Most of the dust falls in spring and summer, so the change of seasons can be seen in the layers of the glacier. In the uppermost layers scientists can see thicker and thinner layers in a pattern that matches their records of the varying amount of snowfall in recent decades. Further down the layers are more compressed but are in agreement with the historical record. Climate changes are detected in glacier layers hundreds of years old that match those from historical accounts such as the "little ice age" in Europe in the Middle Ages. Thick layers of dust are found in glacier layers corresponding to volcanic eruptions documented elsewhere on earth, including Mount Vesuvius nearly 2,000 years ago.

Using similar techniques, scientists count down through the deeper layers, finding evidence for ice ages over tens of thousands of years in the past, in agreement with other geological evidence for the ice ages. The deepest ice core taken from Antarctica went to a depth of two miles, down to the bedrock below the glacier. A count of those layers goes back 720,000 years.

Radiometric Dating

The discovery of radioactivity in the late 1800s and early 1900s led to *radiometric dating* (sometimes called *radioactive dating*), the most precise method of measuring ages. (Carbon-14 dating is one type of radiometric dating.) Radiometric dating can produce inaccurate results when the wrong kinds of rocks are used or incorrect assumptions are made. But the fundamental principles are sound, and consistent results are routinely obtained when careful methods are used. Scientists do not rely on just two or three different radioactive isotopes; they use over forty different radiometric dating techniques, each based on a different radioactive isotope.

Scientists double-check their work by comparing results from different isotopes to make sure that they are consistent with each other. Such double-checking can sometimes be done using multiple isotopes on *the same rock*. Rocks from one formation in western Greenland have had their ages measured more than a dozen times, using five different radioactive isotopes. The results were the same for all five isotopes: an age of 3.6 billion years.

For a detailed explanation of how radioactive decay is used to measure the age of the earth, click on the article "Radiometric Dating" on our website (www.faithaliveresources.org/origins).

Thus geologists today have many lines of evidence that the earth is much older than 10,000 years. Although radiometric dating gives the most *precise* measure of age, it is by no means the only method. The various methods, including some that we do not have space to describe, are independent of each other, relying on completely different techniques and assumptions. Because scientists have multiple independent lines of evidence that have been tested for centuries, the conclusion that the earth is old is scientifically reliable.

MODERN YOUNG-EARTH CREATIONISM

As mentioned earlier, by the early 1900s many conservative Christians, including leaders of the fundamentalist movement, accepted the geological evidence for an old earth. Even so, throughout the late 1800s and early 1900s some Christians maintained a Young-Earth Interpretation of Scripture and tried to reconcile this with the geological data.

Development of Modern Young-Earth Creationism

In 1961 theologian John Whitcomb and engineer Henry Morris published *The Genesis Flood: The Biblical Record and Its Scientific Implications*. This book was particularly influential in revitalizing the modern *Young-Earth Creationism* movement (also called *creation science* or *scientific creationism*). This movement is centered on a Young-Earth Interpretation of Genesis 1, which views the text as historical in the modern scientific sense of history. Proponents believe that God made the world in six *literal* twenty-four-hour days (144 hours) about six to ten thousand years ago.

The young-earth creationism movement also includes a scientific model that claims that modern scientific evidence supports this interpretation. As a scientific model it makes specific predictions: life cannot evolve from nonlife, geological layers will show evidence of recent formation and a global flood, and fossils of plants and animals that are transitional between different life-forms do not exist because each life-form was miraculously created by God.

The movement has research institutes and publishes young-earth creationist textbooks. In public debate young-earth creationists ask to be judged on scientific grounds, claiming that the scientific evidence alone—apart from any religious considerations—fits their model of a recent creation and global flood. In churches, young-earth creationists argue that the Young-Earth Interpretation of Genesis is the best view—or even the only acceptable view—and that scientific evidence supports a young earth. The movement has spread rapidly throughout fundamentalist and evangelical churches, as well as through parachurch organizations in North America.

How did this idea get revived? Geologists in the 1600s started out with exactly the same plan, to find scientific evidence supporting the Young-Earth Interpretation of Genesis 1 and the global flood. As we have seen, this plan was abandoned by 1840 when geologists, many of them Christians, found that the evidence from nature simply did not fit this model. The revival of the Young-Earth Interpretation in recent decades is motivated in part by the same theological concerns that motivated the publication of *The Fundamentals*. It is still a response to liberal Christianity and a rejection of human evolution. But there's a major difference: the original fundamentalists in the early 1900s did not include a recent creation of the earth among the essential Christian beliefs. By contrast, the modern young-earth movement argues that a recent creation is essential to Christian belief.

Another motivation for modern young-earth creationism, according to its leaders, is to respond to atheistic anti-Christian ideas in the public square.

> If the system of flood geology can be established on a sound scientific basis and be effectively promoted and publicized, then the entire evolutionary cosmology, at least in its present neo-Darwinian form, will collapse. This, in turn, would mean that every anti-Christian system and movement (communism, racism, humanism, libertinism, behaviorism, and all the rest) would be deprived of their pseudo-intellectual foundation.
>
> —Henry Morris, *Scientific Creationism,*
> Creation-Life Publishers, 1974.

Proponents of this view argue that an alternative to mainstream science is essential to the fight against atheistic worldviews in our society.

With these goals in mind, the young-earth creationist movement has tried to build scientific arguments in support of a recent creation and global flood. They have responded to scientific data that indicates an old earth by developing alternative models for the data in terms of a young earth.

> For a description of young-earth creationist models and some of the difficulties they face, click on the article "Young-Earth Creationist Views on Continental Drift and Radioactive Dating" on our website (www.faithaliveresources.org/origins).

Besides searching for alternate models to mainstream science, the young-earth creationist movement also searches for some particular features of the earth or the solar system that *prove* that the earth was created recently. Over the past few decades young-earth creationists have published more than 100 such scientific arguments.

THE SHRINKING SUN ARGUMENT

One such argument is the *shrinking sun* argument, used frequently in recent decades and still quoted today by some young-earth creationists. This argument illustrates some of the problems that have affected much of the work in scientific creationism.

In 1979 two mainstream solar astronomers, John Eddy and Aram Boornazian, gave a short presentation at a meeting of the American Astronomical Society ("Secular Decrease in the Solar Diameter, 1836-1953," *Bulletin of the American Astronomical Society* 1979, 11:2). They reported that the sun appeared to have significantly decreased in size over several decades, based on observations of its diameter made at Greenwich Observatory. This was a preliminary result published as an abstract only. It was interesting and compelling enough to share with colleagues at the conference but had not gone through the full peer-review process or been published in a refereed journal.

The result certainly raised the interest and skepticism of other astronomers, since the idea of a shrinking sun didn't fit with the accepted and well-tested model for how the sun works (a model based on nuclear fusion, gravity, pressure, and other known physical processes). If the sun really was shrinking, it could be an exciting indication of new physical forces. Other astronomers immediately set out to check the observations.

In 1980 these three peer-reviewed articles appeared in prestigious journals:

▶ "Observations of a Probable Change in the Solar Radius between 1715 and 1979" (D. W. Dunham, S. Sofia, A. D. Fiala, D. Herald. and P. M. Muller, *Science,* 1980, 210:1243).

▶ "Is the Sun Shrinking?" (Irwin I. Shapiro. *Science,* 1980, 208:51).

▶ "The Constancy of the Solar Diameter Over the Past 250 Years" (J. H. Parkinson, L. V. Morrison, and F. R. Stephenson, *Nature,* 1980, 288:548)

The first article reported measurements of the sun's size based on solar eclipses between 1715 and 1979. The second reported measurements based on transits of Mercury when Mercury passed in front of the sun between 1736 and 1973. These two articles showed that the sun has been basically constant in diameter for at least 250 years. If the sun *were* shrinking, it was doing so much more slowly than Eddy and Boornazian had initially reported. The third article showed that the data used by Eddy and Boornazian was unreliable because of subtle instrumental effects and observational inconsistencies at Greenwich Observatory over the years.

Note that the scientific community did not ignore the challenge of the shrinking sun or deny it because it disagreed with the models. Instead, scientists immediately set out to solve the puzzle. Data from independent sources was used to crosscheck the initial claim, and the original data was checked for problems. Within a year the community had reached a consensus that the sun was *not* shrinking at a dramatic rate. After a few more years of careful study, the sun was found to be oscillating slightly in size, periodically expanding and shrinking once every eighty years.

A few months before the publications in *Science* and *Nature*, Eddy and Boornazian's initial report of a shrinking sun was described to the creation science community by physicist Russell Ackridge ("The Sun Is Shrinking," *Impact* No. 82, Institute for Creation Research, April 1980). Ackridge used the rate of shrinkage reported in Eddy and Boornazian's abstract to calculate that the sun would have been much larger in the past. Assuming that the sun was shrinking at a constant rate, then 100,000 years ago the sun would have been twice its present size, and 22 million years ago it would have been as large as earth's orbit. This obviously is inconsistent with the mainstream scientific claim that the earth is billions of years old. Ackridge concluded, "The discovery that the sun is shrinking may prove to be the downfall of the accepted theory of solar evolution. . . . The entire theoretical description of the evolution of the universe may be at stake."

If the sun truly is shrinking, Ackridge would be right in saying that it would change our understanding of fusion in stars and thus alter our calculations about the ages of stars. But two major problems arise with Ackridge's argument:

▶ First, he used the initial report of shrinkage as though it were well-established fact. But the initial report was not published in a peer-reviewed journal and had not been accepted by other scientists, nor had it been confirmed by other observations. At the time Ackridge made his calculation, scientists in the field had not reached consensus on this topic.

▶ Second, Ackridge assumed that the sun had shrunk at a constant rate throughout its history, ignoring the possibilities of temporary shrinkage or oscillation.

The creation science community began to quote the shrinking sun argument regularly as evidence that the earth could not be billions of years old. They continued to use this argument long after the 1980 papers in *Science* and *Nature* showed that the sun was *not* shrinking. One exception is a short letter by Paul Steidl, published in *Creation Science Quarterly* ("Recent Developments about Solar Neutrinos," 1981, 17:233). Steidl supported the observations by mainstream scientists that the sun was not shrinking. This report went unheeded, as did admonitions from Christian astronomers to look at the newer data (Howard J. Van Till, "The Legend of the Shrinking Sun," *Journal of the American Scientific Affiliation*, 1986, 38, No. 3, pp. 164-74). The "shrinking sun" argument is still quoted today by some young-earth creationists.

For another example of a scientific argument used by the creation science community to support a young earth, see "The Ocean Salt Argument for a Young Earth" on our website (www.faithaliveresources.org/origins).

Creation Science and Scientific Practice

Unfortunately, other creation science arguments have similar flaws. Because there are relatively few young-earth creationist scientists, and because they passionately believe in a young earth and are eager to find evidence for it, some of them neglect good scientific practice. Arguments frequently are accepted and promoted before they have been adequately tested. Young-earth explanations are given for isolated scientific observations at the expense of ignoring a great deal of contradictory data. Some young-earth creationists base arguments on obsolete data and ignore vast quantities of more recent data that tell a different story. They cite a single publication but pay no attention to other publications on the same topic. Meanwhile, they repeatedly accuse the mainstream scientific model of some flaw that scientists have noticed and resolved years ago. Sometimes Christians in these scientific fields attempt to explain the newer data to young-earth creationists, but they are ignored or accused of having an "evolutionary" bias. Sadly, a lack of scientific integrity on the part of some young-earth creationists has tarnished the reputation of the entire young-earth movement. This, in turn, has tarnished the reputation of *all* Christians in the eyes of many scientists.

But others in the young-earth creationist movement act with greater scientific integrity. Some leaders have publicly acknowledged that the shrinking sun argument is flawed. For example, *Answers in Genesis* (www.answersingenesis.org/get-answers/ topic/arguments-we-dont-use) lists it among flawed arguments that creationists should not use. Some creation scientists are aware of a large portion of the current scientific data and actively seek to learn more. They are open to having errors in their models pointed out and avoid repeating arguments that have known flaws. They report the weaknesses as well as the strengths of their own arguments. They try to construct scientific models to explain a large portion of the scientific data rather than just a small, isolated result. They seek to provide an alternative scientific model that explains *all* of the available scientific data in terms of a recent creation. Although they have not succeeded, we should respect young-earth creationists who work with honesty and scientific integrity.

A Young-Earth Interpretation of Genesis does preserve a high view of biblical authority. Those who promote this interpretation typically do so from a desire to have a consistent literal-historical interpretation of passages in the Bible that sound historical. For some Christians this value is strong enough to outweigh the weaknesses in young-earth scientific arguments and the strengths in old-earth scientific arguments.

CONTINUING THE DISCUSSION

Throughout history committed Christians have held many different positions on how to interpret Genesis 1 and 2. In this chapter we've presented four *concordist* interpretations of Genesis: Young Earth, Gap, Day-Age, and Appearance of Age. These views differ in how they interpret both Scripture *and* nature, but each view seeks to reconcile the scientific evidence with the order of events described in Genesis 1. Each interpretation has significant strengths and weaknesses.

But there are still more options to consider. We'll continue this discussion in chapter 6, applying principles of biblical interpretation to Genesis 1 and describing five *non-concordist* interpretations. (You may want to refer back to the chart on p. 100 for an overview.)

QUESTIONS FOR REFLECTION AND DISCUSSION

1. What similarities and differences do you notice between Genesis 1:1-2:3 and Genesis 2:4-25? For example, what methods of creating did God use? What tasks did God give human beings? (See chart on p. 101 for the order of events.)
2. Of the *concordist* interpretations of Genesis presented in this chapter, which had you heard of before? Which were new to you?
3. What did you learn about geology in this chapter? Which of the scientific arguments for age did you find most compelling?
4. What are some signs that a scientific argument is strong? That it is speculative? That it is fundamentally flawed?

ADDITIONAL RESOURCES

More on interpretations of Genesis by the early church fathers:

Catholic Answers. "Creation and Genesis" (www.catholic.com/library/Creation_and_Genesis.asp).

More on geological evidence for great age from a Christian perspective:

Greenberg, Jeffrey. "Geological Framework of an Evolving Creation," *Perspectives on an Evolving Creation.* Keith B. Miller, ed. Grand Rapids, Mich.: Wm. B. Eerdmans, 2003.

Weins, Roger C. "Radiometric Dating: A Christian Perspective" (www.asa3.org/ASA/resources/Wiens.html).

Young, Davis, and Ralph Stearley. *The Bible, Rocks, and Time: Geological Evidence for the Age of the Earth.* Downers Grove, Ill.: Intervarsity Press, 2008.

Young, Davis. "The Discovery of Terrestrial History," in *Portraits of Creation.* Grand Rapids, Mich.: Wm. B. Eerdmans, 1990.

More on the history of the young-earth creationist movement:

Davis, Edward B. "Concordism and American Evangelicals," *Perspectives on an Evolving Creation.* Keith B. Miller, ed. Grand Rapids, Mich.: Wm. B. Eerdmans, 2003.

_____. "Important Primary Texts on Religion and Science in America" (home.messiah.edu/~tdavis/texts.htm).

Numbers, Ronald. *The Creationists.* Berkeley, Calif.: University of California Press, 1992.

More on young-earth creationist views:

Answers in Genesis (www.answersingenesis.org).

Institute for Creation Research (www.icr.org).

More on scientific responses to young-earth creationist arguments from a Christian perspective:
Answers in Creation (www.answersincreation.org).

Reasons to Believe (www.reasons.org).

Ross, Hugh. *Creation and Time.* Colorado Springs: Navpress, 1994.

Van Till, Howard J., Davis A. Young, and Clarence Menninga. *Science Held Hostage.* Downers Grove, Ill.: InterVarsity Press, 1988.

CHAPTER 6

GENESIS: NON-CONCORDIST INTERPRETATIONS

As noted in chapter 2, our life experiences and worldview have an impact on our interests and beliefs about God's world and God's Word. Jim Bradley, a colleague of ours, tells the story of his first serious encounter with Genesis 1:

> I was raised in a nominally Roman Catholic family, pre-Vatican II. Thus I never read the Bible as a child, nor did I attend Sunday school. After graduating from college, I joined the Peace Corps and spent two years teaching high school in India. I was assigned to teach in a school run by Hindu monks, so I was largely with devout Hindus for much of that time. I heard a great deal about their beliefs. Thus, when at the age of 23 I read the Bible for the first time, I was quite struck by Genesis 1. It seemed to me to be quite intentional about speaking to a monotheistic people surrounded by polytheistic cultures. That is, it seemed to me to be systematically going through a list of each of the things I saw my polytheistic friends worshiping, and saying to them, "That's not God. That's a created thing." To me, it was electrifying. In one short chapter, Genesis 1 swept away all of the religious confusion that had surrounded me.

As we continue this discussion of interpretations of Genesis 1, we hope that you will come away with your own "electrifying" moment. Perhaps you're beginning to zero in on a view of your own, or maybe you've held a certain view for a long time. Either way, we trust that your reflection and discussion is causing you to search the Scriptures even more and to stand in awe of the wonders of God's creation.

In chapter 5 we discussed these four *concordist* interpretations of Genesis:

▶ Young Earth
▶ Gap
▶ Day-Age
▶ Appearance of Age

Christians who hold these views believe that God made the earth using the same *sequence* of events described in Genesis 1; interpretations vary in terms of the *length of time* over which the events occurred.

Other Christians, both in the 1800s and today, hold *non-concordist* interpretations of Genesis 1. People who hold these views consider the Genesis text to be divinely inspired and authoritative for the message originally intended. But they do not believe that the text conveys scientific or detailed historical information, at least not in the way that we think of science or historical accounts in the modern world. According to these interpretations, Genesis 1 was not intended to be a textbook of natural history or an eyewitness account. The text tells us about the event of creation as a whole. The sequence and timing of *particular* events described in the text had some cultural or spiritual significance, and the passage as a whole conveys some important theological truths, but the specific sequence of events in that chapter was not intended to be taken as literal, *scientific* truth.

In this chapter we'll discuss five non-concordist interpretations:

▶ Proclamation Day
▶ Creation Poem

▶ Kingdom-Covenant
▶ Temple
▶ Ancient Near Eastern Cosmology

Note that these interpretations have some overlap, and some could be combined without contradiction.

> For a brief summary of both concordist and non-concordist interpretations, see the chart on page 100 in chapter 5.

We'll conclude this chapter by looking at pitfalls in both the concordist and non-concordist interpretations and briefly test the various views against the principles of biblical interpretation introduced in chapter 4.

PROCLAMATION DAY INTERPRETATION

According to this interpretation, which also focuses on the meaning of the days, God took six days to proclaim his creation to his heavenly court. These "days" took place in God's throne room, apart from the time and space of this universe. In other words, on day one God proclaimed in heaven, "Let there be light," and on day five God proclaimed in heaven, "Let the water teem with creatures." The heavenly days have no relation to twenty-four-hour days on earth. The world could have been formed over an entirely different time period, since the Bible teaches that God experiences time differently than we do: "A thousand years in your sight are like a day that has just gone by, or like a watch in the night" (Ps. 90:4; see also 2 Pet. 3:8).

CREATION POEM INTERPRETATION

According to the Creation Poem Interpretation, the days and sequence of events in Genesis 1 were chosen for poetic reasons. Genesis 1 is not typical Hebrew poetry with couplets of repeated or contrasting ideas. But neither is it typical Old Testament historical narrative such as the books of Samuel and Kings and even the later chapters of Genesis. It is a narrative, but a carefully constructed one, including rhythm and repetition of ideas and phrases. Most striking is the structure of the six days. Notice how the first three days are parallel to the second three days. The light sources of day four correspond to light itself on day one, the birds and fish of day five correspond to the sky and sea of day two, and the animals and humans of day six correspond to land on day three. During the first three days the world is formed. It goes from darkness to light and from formless emptiness to clearly defined structures of dry land, ocean, and sky. During the second set of three days, the world is filled with moving creatures: the heavens are filled with lights, the air with birds, the water with fish, and the land with animals and people. Through the six days God works to reverse the initial state of the earth described in Genesis 1:2: "The earth was formless and empty, and darkness was over the surface of the deep waters."

Days of Genesis 1

Initial Problem	Days of Forming	Days of Filling
Darkness	Day 1: Separate light from darkness.	Day 4: Sun, moon, stars
Watery abyss	Day 2: Separate the waters into waters above and waters below, forming firmament and sky.	Day 5: Birds and fish
Formless and empty earth	Day 3: Separate the dry land from the ocean; create plants.	Day 6: Animals and humans

In the ancient world the number seven had special significance, especially in religious texts:

> As regards the seven-day structure, any other temporal order would appear to have been unfitting in that ancient world. Throughout the ancient Near East, the number seven had long served as the primary numerical symbol of fullness/completeness/perfection, and the seven-day cycle was an old and well-established convention. . . . [It] added symbolic reinforcement of the explicit themes of the completeness of God's creative work and the "goodness" of the created realm.
>
> —John H. Stek, "What Says the Scripture?" *Portraits of Creation*, 1990.

The inspired human author crafted the text of Genesis 1 to convey the goodness of God's creation, its completeness (all aspects of existence, both structures and moving creatures), and its orderliness (in contrast to its initial dark chaos). The careful structure of this passage shows that the author selected the sequence of events and the number of days with symbolism and thematic order in mind rather than according to our modern scientific concept of historical sequence. The organization and structure of the text support non-concordist interpretations of Genesis 1, since it appears that historical sequence was not the top priority of the original author.

KINGDOM-COVENANT INTERPRETATION

People of the ancient Near East were also familiar with land grant covenants and suzerains. (*Suzerains* were powerful rulers who gave limited authority over a region to their vassals.) Supporters of the Kingdom-Covenant Interpretation point out that Genesis 1 shares these themes. God, the great King, creates the realms of his kingdom on the first three days. During the second set of three days, God populates them with creatures who serve him and each other. Genesis 1:24-29 calls special attention to the role of humans: they are the special covenant vassals of the great divine Suzerain. Humans bear the King's image and are granted limited authority. Thus the message of the text is not primarily about the timing or the formation of physical structures but about setting up the relationships among God, nature, and humanity.

TEMPLE INTERPRETATION

The Temple Interpretation was recently developed by John Walton, a professor of Old Testament at Wheaton College. In the ancient Near East a temple was viewed as a mini version of the whole cosmos. The Bible turns this around to portray the cosmos as God's temple (Isa. 66:1-2). Thus Genesis 1 can be read as the inauguration of God's temple, with six days of preparation culminating in God taking up residence ("rest") on day seven. Walton emphasizes that these are not days of creating the physical material of the universe but days of establishing the *function* of each piece out of the original disorder. Genesis 1 does not speak about the timing and formation of the material universe and thus does not conflict with whatever science might have to say about natural history.

ANCIENT NEAR EASTERN COSMOLOGY

The Ancient Near Eastern Cosmology Interpretation emphasizes that Genesis 1 teaches a profound *theological* message: that God created everything. This message is embedded in the ancient Near Eastern *physical* picture of the cosmos.

Consider the cultural and historical context of Genesis 1. Archeology has taught us much about ancient cultures over the last few centuries. While scientists have been studying the earth, biblical scholars and archeologists have been learning more about the ancient Near Eastern nations surrounding the Hebrews. The ancient Hebrews were neighbors to—and sometimes exiles among—the peoples of Egypt, Babylonia, and Canaan. The Hebrews would have been aware of how those cultures viewed the cosmos and its origin, and they would have been thinking about these views when they heard the Genesis text. To figure out what Genesis meant to the Hebrew audience, we need to understand these surrounding cultures.

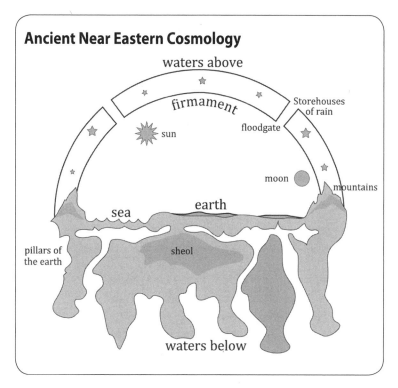

Ancient Near Eastern Cosmology

waters above

firmament

sun

Storehouses of rain

floodgate

moon

mountains

sea

earth

pillars of the earth

sheol

waters below

The ancient Near Eastern view of the world is very different from our modern scientific understanding of the world. Archeology tells us that ancient peoples did not believe in a spherical globe. They believed that the earth was flat, surrounded by water above and below. They also believed that the sky was a solid dome or *firmament* that held back the ocean of primeval waters above the sky. The idea of a firmament seems bizarre to us now, but it was a way for a pre-scientific culture to make sense of the world around them. Without any understanding of evaporation and precipitation, they simply concluded that an ocean of water existed above the sky. A solid dome in this view held the waters up, and occasionally floodgates or windows would open and allow rain from the storehouses to fall to the earth. They also believed that an abyss of primeval waters was under the earth; this fit with their experience of digging wells and seeing springs of water rising up out of the ground. Without any understanding of the earth rotating on its axis, they assumed that the sun literally traveled from east to west each day, just as it appears to do, and that it traveled under the earth each night to return to the east.

This physical picture of the world matches the first three days in Genesis 1. Starting from a dark, formless, watery abyss, God separates light from darkness on day one. On day two God separates the waters into "waters above," which are held back by the solid firmament, and "waters below," with the sky in between. On day three God gathers the waters below the sky, sets boundaries on them, and forms the dry land. (This same sequence appears in Ps. 104:1-9.)

The "firmament" on day two is designed in various ways, according to different English translations. See www.biblegateway.com, or "Genesis 1:6-8 from Five English Translations" on our website (www.faithaliveresources.org/origins).

For a discussion of other Bible passages that refer to the physical structure of ancient Near Eastern cosmology, see "Ancient Near Eastern Cosmology in the Bible" on our website (www.faithaliveresources.org/origins).

Egyptian Engraving of the Cosmos

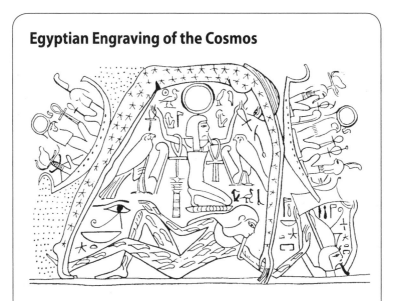

In this highly symbolic Egyptian representation of the world, the starry sky (the lady of heaven—the goddess Nut) arches over the reclining earth (the god Geb). Above her is the upper ocean in which the solar barque [boat] sails to carry the sun (the falcon-headed god Re) from the eastern horizon up to the zenith and then down to the western horizon. (Re is accompanied by the goddess Maat with her identifying feather; she is the daughter of Re, who is the source of world order.) Kneeling above the earth (reclining Geb) and holding up the sky (Nut) is the atmosphere god Shu; he holds in both hands the symbol of the breath of life. At the lower right is Osiris, the great god of the world of the dead. Not represented here is the subterranean ocean which the solar barque traverses at night to return the sun at dawn to the eastern horizon. That ocean is clearly shown in other symbolic representations.

—Engraving from Othmar Keel, *The Symbolism of the Biblical World: Ancient Near Eastern Iconography and the Book of Psalms,* Eisenbrauns (June 1997); scanned from *Portraits of Creation.* Explanation from John H. Stek, *Portraits of Creation,* Wm. B. Eerdmans, 1990.

This physical picture of the world is woven throughout the Old Testament. It reflects the way the ancient Egyptians, Babylonians, Canaanites, and Israelites *literally* understood the world around them. In the account of Noah's flood, the waters come both from the floodgates of heaven and from the springs of the deep (Gen. 7:11), not from clouds. In Psalm 19:4-6, the sun rises at one end of the heavens and follows its path to the other end, just as the Egyptians viewed the sun as a boat sailing on the firmament. Thus, if Christians today wish to interpret Genesis 1 in a completely literal way, they must believe that

▶ the earth is flat rather than spherical.

▶ the sky is a solid dome rather than a transparent atmosphere.

▶ the sun, moon, and stars move along the sky dome around the earth.

▶ there is an ocean of water is above the sky.

Any other so-called *literal* interpretation of Genesis 1 is, at best, a semi-literal interpretation in which the reader picks and chooses some pieces to view literally and others to view figuratively.

Babylonian Picture of the World

This physical picture of the world is also evident in the Babylonian epoch story *Enuma Elish*, which tells how the Babylonians believed the world came to be. The story begins with primeval chaos and then tells of the origin of light, a sky-dome, the dry land, and a means of marking time. Each of these physical structures is viewed as either a god itself or as closely related to the gods. The story ends with the gods resting and celebrating in a banquet. (The complete text can be found in James Pritchard, *Ancient Near East Texts Relating to the Old Testament*, 3rd edition, Princeton University Press, 1969. An English translation is also available at www.cresourcei.org/enumaelish.html.)

Given that Genesis 1 and the Egyptian and Babylonian cosmologies share a similar physical picture of the world, the differences among them are easy to spot. In both the Egyptian engraving and the *Enuma Elish* story, a pantheon of gods is present, with gods inhabiting various parts of the physical world. In *Enuma Elish* matter is coeternal with the first gods. From the first pair of gods (Apsu and Tiamat) six generations of gods are born. Ultimately Tiamat is defeated by the god Marduk, who splits Tiamat's body like a clamshell, using half to form the sky dome. Marduk also sets the lesser gods in "stations" to mark days, months, and seasons. At the end of this story humans are made from the flesh of a defeated god in order to be slaves to the gods and servants at the banquet.

Genesis 1 Versus *Enuma Elish*

Genesis 1	Enuma Elish
There is one God.	There is a pantheon of gods.
God created an ordered world by the authority of his Word.	The world was formed as the result of battles among the gods.
No part of the physical world is divine.	The sun and moon and other physical objects are gods that control the fate of humans.
God declared all parts of creation good.	Some physical structures are related to good gods, others to bad gods.
God created men and women in his image as the culmination of the story and declared them very good. God gave them responsibility to be stewards of creation.	Humans are made as an afterthought at the end of the story from the flesh of a defeated god, to be slaves to the gods.

Imagine yourself as one of the ancient Israelites hearing the Genesis 1 account of creation for the first time. Abraham and Sarah, your founding parents, came from Mesopotamia. Your ancestors were slaves in Egypt who settled in Canaan. Many details of Genesis 1 sound familiar because you've heard the stories from the surrounding cultures. You've heard about the dark primeval chaos at the beginning, the formation of a sky dome, the sun and moon moving through the heavens to mark the days and seasons, and the origin of humans.

But wait! This creation story is different from those of Egypt and Mesopotamia! No gods are mentioned—only the one *God*. The sun is not the powerful Egyptian god Re, nor even a minor god subject to other gods. The sun is not a god at all but a physical object, the brighter of two lights that God made. Instead of the body of a god such as Tiamat or Nut holding up the waters above, the firmament is a physical, non-divine object that God made and named. In fact, *none* of the living things and physical structures in the story are divine; all are made by the authority of the one and only God. Israel's God is God of *every* part of the cosmos—an astonishingly bold claim compared to the other creation stories told by your neighbors. The more you hear, the more striking are the differences between your God and the gods in the Babylonian story (see chart above).

Because of this context the original audience would not have heard Genesis 1 teach that the earth was formed out of a watery chaos or that there was a solid dome firmament holding back waters above the sky. They already believed that physical picture! Rather, the original audience heard Genesis 1 as a powerful *theological* manifesto proclaiming the true authority of the God of the Israelites and the true status of humanity. God inspired the human author of Genesis 1 to communicate these theological truths using a physical description of the earth that was familiar to them. Imagine that God had instead tried to correct their scientific misconceptions by explaining to them that the earth is spherical (not flat) and the sky is gaseous (not a solid dome), and that it formed over billions of years (longer than they could comprehend). It would have baffled them completely! Moreover, it would have completely distracted them from the theological message. God graciously accommodated himself to the needs of the people at that time by communicating the spiritual message in the clearest means possible, rather than obscuring it within scientific information. Biblical scholar Dan Harlow writes:

> Genesis 1 tells us nothing factual about the age or size of the universe, about the physical processes by which either the earth or life on earth developed, or about the order in which different forms of life emerged on our planet. Instead, it affirms the sovereignty of God, the goodness of creation, and the dignity of humanity.

These theological truths are timeless and normative for us, but the ancient cosmology that serves as their vehicle is not.

> —"Creation According to Genesis: Literary Genre, Cultural Context and Theological Truth," *Christian Scholars Review*, 2007, p. 181.

COMPARING INTERPRETATIONS

In chapters 5 and 6 we've presented nine different interpretations of Genesis 1. How should Christians go about choosing among all of these interpretations? Such a decision should be based on consistent principles and prayerful reflection, not just on "what sounds good." Here are our own conclusions.

Weaknesses in Concordist and Non-Concordist Interpretations

Both concordist and non-concordist interpretations of Genesis 1 arise from good motives, a desire to show that the Bible does not conflict with nature's testimony. But both types of interpretations have their pitfalls.

For concordists, the temptation is to interpret every Bible verse to match the current scientific picture. The meanings of particular phrases can be bent out of shape to match a particular scientific finding. For example, Hebrew words that literally meant *birds* or *plants* to the original audience are redefined to meet some modern scientific category such as insects or single-celled organisms, just to make the order of events line up. By focusing on trying to match the details of the ancient text to twenty-first century knowledge, the concordist may miss meanings in the passage that were clear in the original cultural context, including important spiritual insights. Moreover, concordists can be forced to regularly change and update their interpretations as modern scientific knowledge grows and changes. For instance, the Gap

Interpretation twisted the meaning of Genesis 1:2 outside its original intent; later it failed to match new scientific evidence.

For non-concordists the temptation is to interpret every Bible verse that appears to disagree with science as figurative without first studying the text. By interpreting a text that was intended to be understood literally as metaphoric, they may bend the meanings of particular phrases to refer to purely spiritual ideas and ignore the historical meanings they had in the original cultural context. At one extreme non-concordists can apply the same strategy to all Bible passages and even interpret Jesus' miracles and resurrection as spiritual symbols simply because they think that miracles are scientifically impossible.

For both concordists and non-concordists the temptation is to let science drive the interpretation of Scripture more than it should. When an apparent conflict arises between science and a biblical text, it can and should motivate us to consider a biblical passage more closely. The scientifically discerned testimony from God's book of nature can even be a useful tool for deciding between two or more biblical interpretations that are otherwise equally valid. But the interpretations themselves are not *determined* by science; they must be driven by theological considerations and be consistent with the rest of Scripture.

To avoid these risks we need to look at what the best biblical scholarship has to say about the passage rather than at how it fits with science. Finally, we must take care that the desire to resolve conflicts does not distract us from the main message God has for us in the text. Our primary calling as Christians is to live our lives according to the clear messages of God's Word; it is a lesser calling to debate the subtleties of interpretation of less clear passages.

Genesis 1 in Its Original Context

To choose among the various interpretations, we recommend using a consistent approach based on the principles of biblical interpretation discussed in chapter 4 (p. 84). The first principle, that each passage should be interpreted in light of the rest of the Bible, provides some guidance. For instance, the Bible's teaching on God's truthfulness and his glory displayed in creation might lead us away from the Appearance of Age Interpretation. The differences between the Genesis 1 and Genesis 2 accounts might point toward a non-concordist interpretation.

The second principle of interpretation gives more direction. It reminds us *first* to work out what the passage meant in its original literary, cultural, and historical context, and *then* figure out what meaning it has for us today. How do the various interpretations fit this principle? Of the four *concordist* interpretations discussed in chapter 5, the Young-Earth Interpretation seems to come closest to what ancient peoples would have heard in the text. The Gap and Day-Age concordist views would have baffled the original audience, since these ancients would have had no concept of geological ages; if they could not fathom time periods of millions or billions of years, the text must have meant something different to them.

Of the five *non-concordist* interpretations of Genesis discussed in this chapter, the Proclamation Day Interpretation, while it has some basis in the text, seems least likely to be the meaning heard by the original audience. The proclamations are implemented as soon as God says them, and there is no reference to a different timing or sequence of events in terrestrial time. In our view a combination of the Ancient Near Eastern Cosmology, Kingdom-Covenant, Temple, and Creation Poem Interpretations come closest to what the original audience would have heard. The differences between the Genesis text and the pagan stories highlight the sovereignty of God and the goodness of creation. The elegant poetic structure and inspired phrases reinforce the theological messages of the Kingdom and Temple interpretations.

Genesis 1 for Modern Readers

With a better understanding of what the original audience heard, we have insight into God's message for them and thus for us. *If God's purposes in Genesis 1 did not include teaching scientific facts to the Israelites, then we should not look here for scientific information about the age or development of the world.* For modern readers, as for the original audience, the message of this chapter is its powerful theological truths. God does not use the Bible to teach us the physical processes he uses to make the rain fall or the earth orbit the sun or to form the mountains. Instead, in a beautifully crafted and impressively short text, God teaches us all about

- his sovereignty.
- the goodness of creation.
- the honored status of humankind as his imagebearers.

God has given us a text that speaks of the physical world in simple terms, based on how it appears, in order that all people might understand it. This is not a modern idea. Centuries ago a theologian wrote:

> For to my mind this is a certain principle, that nothing is here treated of but the visible form of the world. He who would learn astronomy and the other recondite arts, let him go elsewhere. Here the Spirit of God would teach all men without exception . . . It must be remembered, that Moses does not speak with philosophical acuteness on occult mysteries, but states those things which are everywhere observed, even by the uncultivated, and which are in common use.
>
> —John Calvin, *Commentaries on The Book of Genesis.*

The common language of this text has made it accessible to people of many times and cultures, aiding the communication of the gospel around the world.

Does a non-concordist interpretation of Genesis 1 mean that we have sacrificed a literal understanding of the gospel? No. The Gospels were surely heard by their first audience as historical eyewitness accounts by the disciples, and everything about the emphasis and tone in those books indicates that Jesus' resurrection and miracles are essential events in the story. That is how we should read the Gospel stories still today. In Genesis 1, on the other hand, the first listeners heard nothing new about the physical universe; all the emphasis was on *who* created the world and humanity and *why* they were created.

What does this mean for science? It means that Genesis 1 is not a science textbook. The text was never intended to teach scientific information about the structure, age, or natural history of the world. Thus, comparing Genesis 1 to modern science is like comparing apples to oranges. Or perhaps more accurately, comparing Genesis 1 to modern science is like comparing Psalm 93:1 ("The world is firmly established; it cannot be moved") to modern astronomy. Genesis is neither in agreement *nor* in conflict with the sequence of events found by astronomy and geology.

As scientific knowledge increases and changes over the centuries, its understanding of the physical structure and history of the earth will change. But through all of those centuries the theological truths of Genesis 1 remain the same: there is one sovereign God who makes light from darkness, creates an ordered world from chaos, and fills an empty world with good creatures. Humans need not fear the capricious whims of a pantheon of gods but can instead trust in the one true God who made us in his image and declares us "very good."

To close this chapter, we quote from *Our World Belongs to God,* a testimony written a few decades ago. It is a beautiful statement of central Christian beliefs about creation.

Our world belongs to God—
not to us or earthly powers,
not to demons, fate, or chance.
The earth is the Lord's.

In the beginning, God—
Father, Word, and Spirit—
called this world into being
out of nothing,
and gave it shape and order.

God formed sky, land, and sea;
stars above, moon and sun,

making a world of color, beauty, and variety—
a fitting home for plants and animals, and us—
a place to work and play,
worship and wonder,
love and laugh.
God rested
and gave us rest.
In the beginning
everything was very good.

Made in God's image
to live in loving communion with our Maker,
we are appointed earthkeepers and caretakers
to tend the earth, enjoy it,
and love our neighbors.
God uses our skills
for the unfolding and well-being of his world
so that creation and all who live in it may flourish.
—*Our World Belongs to God,* paragraphs 7-10.

QUESTIONS FOR REFLECTION AND DISCUSSION

1. Numerous Bible passages besides Genesis refer to ancient Near Eastern cosmology. Look up Exodus 20:4; Psalm 93; and Proverbs 8:22-29 in various Bible translations. (Several translations are available at www.biblegateway.com.) Read them in context. What words or phrases fit with the ancient Near Eastern picture?

   These and several other passages are discussed briefly on our website (www.faithaliveresources.org/origins). Click on "Ancient Near Eastern Cosmology in the Bible."
2. Which interpretation or interpretations of Genesis do you think are best? Why? What weaknesses do these interpretations have, and how do you handle them?

ADDITIONAL RESOURCES

Collins, John C. *Genesis 1-4: A Linguistic, Literary, and Theological Commentary.* P&R Publishing, 2006.

Enns, Peter. *Inspiration and Incarnation: Evangelicals and the Problem of the Old Testament.* Grand Rapids, Mich.: Baker Academic, 2005.

Glover, Gordon. *Beyond the Firmament: Understanding Science and the Theology of Creation.* Watertree Press, 2007.

Hyers, Conrad. "Comparing Biblical and Scientific Maps of Origins," *Perspectives on an Evolving Creation,* Keith Miller, ed. Grand Rapids, Mich.: Wm. B. Eerdmans, 2003.

Stek, John H. "What Says the Scripture?" *Portraits of Creation.* Grand Rapids, Mich.: Wm. B. Eerdmans, 1990.

Walton, John H. *The Lost World of Genesis One: Ancient Cosmology and the Origins Debate.* Downers Grove, Ill.: Intervarsity Press, 2009.

CHAPTER 7

AN ANCIENT AND DYNAMIC UNIVERSE

H ave you ever seen the Milky Way? Many people today have not. The lights in modern cities overshadow all but a few dozen stars. But if you've lived in a rural area or been camping in a remote place, you've seen a true dark sky with thousands of brilliant stars on a deep black background. Among the bright constellations the Milky Way is often visible as a silvery streak across the sky, the light coming from millions of stars in our own galaxy. Throughout most of history people saw a dark sky like this every clear night. Your great-grandparents saw it, and Abraham and Sarah saw it. When the psalmist saw it he sang, "The heavens declare the glory of God, the skies proclaim the work of his hands" (Ps. 19:1).

Astronomy today is much more than star-gazing. It is one of the most exciting areas of twenty-first century research, with high-tech telescopes located on every continent and in space. Information about the universe is continually arriving on earth in the form of visible light, radio waves, microwaves, and x-rays. By analyzing all this data, astronomers have found that the universe

▶ is incredibly huge, filled with billions of galaxies that each contain billions of stars.

▶ has a long and dynamic history with a dramatic beginning.

▶ is old, but not infinitely old.

These are wonderful, awe-inspiring scientific discoveries. For Christians they are more than just cool science—they also declare God's glory. Modern telescopes and space probes allow us to see much more of the heavens than Abraham or Sarah or David could see. They show us more and more examples of the beauty, power, and careful handiwork of God's creation. Just as a painting declares something of the character and ability of the artist, so the universe teaches us something of the character and power of its Creator. In this chapter we will describe these discoveries and what they may be declaring about God.

THE UNIVERSE IS VAST

Imagine standing on a street corner at night looking at a parking lot filled with streetlights. How could you figure out the distance to each light? The easiest way would be to use a measuring tape from one to the next. But if you weren't able to move from the street corner, could you still figure out the distance? This is the situation astronomers face. Astronomers want to know how far away the stars and galaxies are from earth, but they can't travel to the stars to measure the distance directly.

Measuring Distance by Motion

One way you can find the distance to the streetlights is by using *depth perception*. With two eyes your brain can compare the images seen by each eye to estimate the distance to the nearest streetlights. For the more distant streetlights you can improve your depth perception by stepping from side to side while looking at the streetlights. As you move back and forth, nearby lights appear to move back and forth more than distant lights. Then, using just a few facts about triangles, you can calculate the distance to the nearer streetlights.

Astronomers use a similar technique, called *parallax*. Parallax is an optical illusion in which a nearby stationary object appears to move because of the motion of the observer. (For more information, click on "Parallax and Its Role in the Heliocentric/Geocentric Debate" at our website, www.faithaliveresources.org/origins.) Astronomers create the illusion by waiting for the whole earth to move side to side in its orbit around the sun and then by comparing observations made six months apart. Nearby stars appear to move more than do distant stars. Using parallax, astronomers have measured the distances to all the stars within 3,000 light-years of Earth. This method doesn't work on more distant stars because their apparent motion is too slight to measure.

What's a Light Year?

Light travels at a speed of 186 thousand miles per second. That means light takes 1.2 seconds to travel the 238,000 miles between the Moon and Earth. Light takes 8.3 minutes to travel from the Sun to Earth. Light takes about five hours to travel between Earth and Pluto. The nearest star is much farther away from Earth, and light takes 4.3 *years* to travel from it to Earth. Distances in the universe are so vast that astronomers do not measure in miles but in *light years*, the distance light travels in one year. Thus, the nearest star is 4.3 light years (about 25,000,000,000,000 miles) away.

Measuring Distance by Brightness

Another way to measure the distance to the streetlights is by their brightness. If someone asks "Which streetlight is the brightest?" you'll answer "The nearest one." You know that the fainter lights must be farther away. Astronomers use the same method. For two otherwise identical stars, the nearby star is brighter and the distant star is fainter. Astronomers measure the brightness carefully and use a mathematical relationship to calculate the distance to the stars. But this method won't work unless the stars each have the same energy output. Astronomers call the light energy output of stars their *luminosity*. If some streetlights have 300-watt bulbs and others 1000-watt bulbs, the 1000-watt bulbs will look brighter and can trick you into thinking they are closer. To make this method work, astronomers need a standard *candle*, a type of object that always emits the same amount of light.

It can be hard to find a standard candle. You may have noticed that some streetlights are a yellow color while others are a brighter blue-white color. The colors are due to the type of gas in the bulb; the yellow ones use sodium, while the blue-white ones use mercury. You may also have noticed that the sodium lights don't put out as much light as do the mercury lights. If you were to assume that all streetlights have the same luminosity, you'd get inaccurate results for distance. Astronomers run into the same kind of problem: stars are not identical, and some emit more light than others.

You can look at the color of the light from a streetlight to determine whether it's sodium or mercury. Then, using the correct value for each light's luminosity, you can calculate the distance correctly. In effect you are separating the lights into two different standard candles. In the same way, astronomers use the color of a star and other information contained in the spectrum of its light to tell what kind of star it is and how much light it produces. Once they've identified the type of star and know its luminosity, it's easy to measure the brightness and calculate the distance to the star. Using this method on stars and star clusters throughout our galaxy, astronomers have found our Milky Way galaxy to be 150,000 light years across.

One particularly useful standard candle is the "Cepheid variable star." *Cepheids* (pronounced *SEF-ee-id*) are thousands of times more luminous than the Sun, making them easy for a telescope to spot at great distances. Cepheids are also easy to distinguish from other stars because they emit brighter and fainter light in a periodic pattern; some pulse every few weeks and others every few days. The *length* of the pulsing pattern turns out to be directly related to average *luminosity*; longer pulses correspond to greater luminosity. This means Cepheids are excellent standard candles because astronomers can simply measure the time between pulses to figure out the luminosity. Once the luminosity is known, it's easy to measure the brightness and calculate the distance to the star, and thus the distance to the galaxy that it's in. Astronomers have used this method to precisely measure distances to galaxies up to 70 million light years away.

Astronomers have several other useful standard candles. The fact that astronomers have multiple methods to measure distances allows them to crosscheck their results and catch problems. Some methods calculate the distance to nearby stars, others to faraway stars, and still others to galaxies near and far. One of the standard candles, a certain type of supernova explosion, has been used to measure the distances to galaxies as far as 10 *billion* light years away.

Measuring Distance by Apparent Size

When standing on the street corner, you can also note how tall the poles are. If someone asks "Which street light looks the tallest?" you'll answer "The nearest one." Nearby posts look taller, and more distant ones look shorter, even though they are all the same physical size. Astronomers use the very same method with astronomical objects like galaxies and star clusters. When powerful telescopes look deep into the universe, they detect galaxies that appear extremely tiny because they are located billions of light years away.

Where do you live? Perhaps you're thinking of your street or neighborhood, your city, or even your country on Earth. But where do you live in relation to the Universe? Earth, of course, is not the only planet, but a member of a small neighborhood of planets orbiting the Sun, called the Solar System. The Solar System is only one neighborhood in a vast city of billions of stars called the Milky Way Galaxy. The Milky Way Galaxy is not alone in space but is one of a few dozen such cities on an island called the Local Group. And the Local Group is itself one of the smaller islands in an island chain of galaxy groups and clusters nearly a billion light years long called the Virgo Supercluster. The Universe is filled with supercluster chains like this one. Thus, your entire address would be 123 Any Street, My City, Our Country, Planet Earth, Solar System, Milky Way Galaxy, Local Group, Virgo Supercluster, the Universe.

The following interactive websites allow you to explore the cosmos from the solar system out to the largest superclusters:

▶ "An Atlas of the Universe" from Richard Powell. www.atlasoftheuniverse.com/.

▶ "Secret Worlds: The Universe Within" from Molecular Expressions. micro.magnet.fsu.edu/powersof10/.

▶ "A Question of Scale" from Bruce Bryson. www.wordwizz.com/pwrsof10.htm.

Everyday Comparisons

It is difficult to comprehend how large the universe really is, but some comparisons using everyday objects can help. Imagine shrinking the solar system down until the Sun is the size of a tennis ball. On that scale Earth would be the size of the period at the end of this sentence, orbiting about 20 feet away from the tennis ball. Pluto would be a much smaller dot, orbiting nearly three football fields away! The nearest star would be another tennis ball located 1,100 *miles* away.

To get a feel for the Milky Way galaxy, we have to shrink the model again. Take the whole solar system (including the three football fields out to Pluto) and shrink it to the size of the tiny ball tip on a ballpoint pen. On this scale, the galaxy would be a huge disk, 150 miles across, and the tiny ball of our solar system would be located 28 miles from the center.

The Milky Way is just one galaxy in a universe filled with more than 10 billion galaxies. Each of these galaxies contains billions of stars. The astronomical extent of the universe is so vast that it can make our lives on the earth seem incredibly small and insignificant. The atheist astronomer Carl Sagan described the size of the universe this way:

> Our planet is a lonely speck in the great enveloping cosmic dark. In our obscurity, in all this vastness, there is no hint that help will come from elsewhere to save us from ourselves.
>
> —*The Pale Blue Dot,* Random House, 1994.

Implicit in the response of atheists like Sagan is the view that God is merely an idea invented by humanity. The universe is huge, they reason, and humanity is small, so the God invented by humanity must also be small and irrelevant to the universe. If this is so, Sagan could rightly conclude (above) that "no . . . help will come from elsewhere to save us from ourselves." But the Bible declares that God came before humans and that God rules the whole cosmos, not just the earth. The scientific discovery of the vastness of the universe does not contradict this belief but instead strengthens it by demonstrating how large God's reign really is. The vastness of the universe challenges the idea of a small God who is only relevant to our earthly concerns. It stretches our imaginations to glimpse the true God, who is as far beyond us as the distant galaxies are beyond earth.

Why Us?

How could this powerful God, governing such a vast universe, care about us? Is it audacious to believe that God loves us when we are such an infinitesimally small component of what he has made? This is not a new question. Over three thousand years ago the psalmist looked up at the night sky and wrote,

LORD, our Lord,
> how majestic is your name in all the earth!
When I consider your heavens, the work of your fingers,
> the moon and the stars, which you have set in place,
what are mere mortals that you are mindful of them,
> human beings that you care for them?
—Psalm 8:1, 3-4.

It is the same question that is asked today. The psalmist goes on to answer the question in verses 5-8:

You have made them a little lower than the heavenly beings
> and crowned them with glory and honor.
You have made them rulers over the works of your hands;
> you put everything under their feet:
all flocks and herds,
> and the animals of the wild,
the birds in the sky, and the fish in the sea,
> all that swim the paths of the seas.

God gave humanity a significant role to play in the cosmos, that of caring for our planet.

Psalm 8 is not the only Bible passage that directs our thoughts to the vast cosmos. In Psalm 103:10-12 we read that

[God] does not treat us as our sins deserve
> or repay us according to our iniquities.
For as high as the heavens are above the earth,
> so great is his love for those who fear him;
as far as the east is from the west,
> so far has he removed our transgressions from us.

How far is east from west? To our modern ears this can sound like a four-hour plane flight from coast to coast. But recall how ancient Near Eastern cultures like that of the Hebrews viewed the cosmos. They believed that the full extent of the cosmos was from the flat earth up to the hard dome of heaven and from the eastern horizon to the western horizon. To their ears these verses refer to the *entire created universe*, not just to some part of planet Earth. This is the object lesson the psalmist used to illustrate the love of God for humanity. God's love is as vast as the cosmos itself, and God's forgiveness removes our sins to the end of the universe! This psalm clearly is not teaching us to view the vastness of the cosmos as a sign of how small we are in God's eyes but rather as a sign of the greatness of God's love. Our significance is not based on our relative size in the universe but on the demonstrated love of God for each one of us.

THE UNIVERSE IS DYNAMIC

A century ago astronomers thought the universe was static. They thought that the basic properties of the universe and of the stars and galaxies in it were always the same and did not change significantly over time. During the last century we've learned that this assumption was completely wrong. The universe and the objects in it have undergone dramatic change and development since the beginning of the universe.

Dynamic Change in Stars

Stars are a primary example of this ongoing change. Stars are not the same forever; each goes through a complete life cycle. Since stars live much too long for astronomers to watch a single star go through all its stages, they study many stars at all stages of their lives. They look at stars large and small, hot and cold, young and old, and use the laws of physics to work out how an individual star changes over time.

The scientific model of a stellar life cycle is very solid, explaining many observations and successfully predicting the properties of new stars and star clusters that are discovered. A star changes as follows over its life cycle:

▶ **Star Life.** A typical star, like the Sun, is stable for most of its life, similar to the adult stage of a plant or animal. This stage is very long, typically billions of years. During this time the star steadily converts hydrogen into helium in its core, but its appearance on the outside hardly changes.

▶ **Star Aging.** When all of the hydrogen in the core is used up, the star begins to change dramatically. It balloons out into a huge red giant star, as big as the orbit of the earth around the Sun. It then goes through a few shorter stages, during which it produces many of the different types of atoms with which we are familiar, including carbon, nitrogen, and oxygen.

▶ **Star Death.** A star like the Sun dies gradually as its outer layers drift away from the core, leaving behind a hot white dwarf star. Stars that are bigger than the Sun die much more dramatically in a supernova explosion (see photo on p. 64). In a tiny fraction of a second, most of the star's material is blown outward in a powerful burst of light, energy, and particles. A supernova can release so much energy that for a while it outshines a whole galaxy! It is during this stage that the helium, carbon, nitrogen, and oxygen atoms produced in the star get spread out through space, enriching the thin cloud of gas and dust between the stars.

▶ **Star Birth.** New stars are formed inside these rich clouds of gas and dust (they are called nebulae; see photo on next page). While they do not descend from previous generations the way plants and animals do, they do incorporate the atoms produced in the previous generation of stars. Many stars form at about the same time in a single nebula, producing a star cluster in which all the stars are about the same age. On average, one new star per year is born in our galaxy.

The interstellar nebula NGC346 is a dynamic region of space about one million times larger than our solar system. It is an active stellar "nursery" in which new stars are forming within clouds of gas and dust. The bright light from the stars causes the surrounding gas to glow in intricate filaments and beautiful swaths of light. For a dramatic color version of this Hubble Space Telescope photo, visit *http://hubblesite.org/newscenter/2005/35/*.

▶ **Planet Birth.** Planets are born alongside their parent star. Dust grains and atoms from the nebula form a disk of material swirling around the star. In the disk dust grains gradually clump together. The clumps collide and grow in size from pebbles to boulders to asteroid-size to planets. While the details of planet formation are not completely understood, astronomers have detected this process happening in dusty disks around nearby stars.

> For a deeper scientific understanding of a star's life cycle and planet formation, check out "The Life Cycle of Stars" on our website (www.faithaliveresources.org/origins).

The dynamic life cycle of stars illustrates the constant change and development happening throughout our galaxy and universe. These are the natural processes God used to create our own planet. The atoms that make up earth and our bodies were formed in an earlier generation of stars long ago. Those stars exploded, scattering their atoms throughout the nebulae. In one of these nebulae a new star formed with a dusty disk swirling around it. From that disk Earth and other planets formed. Thus the oxygen we breathe into our lungs was once part of a glowing nebula between the stars. A Christian might say "God made us from stardust."

Creation and Providence

The life cycle of stars is one example of God's ongoing creation of new things. Nature is telling us that God chose not to make each individual star miraculously at the beginning of the universe. Instead God designed a dynamic process by which generations of stars are born, grow old, and die throughout the history of the universe.

In everyday language Christians say that God *creates* new things like stars, trees, butterflies, and so on. God's ongoing creation of new things is intimately tied to his *providence*. God upholds the laws of physics, by which new stars form over long periods of time. And God upholds the laws of chemistry and physics, by which trees grow from seeds and caterpillars turn into butterflies.

> Theologians sometimes make a distinction between God's work of *creation* and his work of *providence*. In theological language God's work of creation is completed, and he is now resting from that work (Gen. 2:1-3). God's ongoing governance of the universe falls under the category of *providence*.

In one sense God's work of creation was completed with the initial creation of matter, energy, time, and space at the beginning of the universe. But as God has providentially governed the universe since then, the forms taken by matter and energy have changed. Stars, galaxies, and the universe itself are not static and unchanging but are vibrantly dynamic. In fact, the book of nature is telling us that most things, including the sun and the earth, formed well after that initial creation event.

We could say that God is continually *creating* new stars. Or we could say that God is *providentially* maintaining the natural laws of gravity, pressure, and nuclear fusion through which new stars are constantly forming. Theologically these two sentences might sound different, but they refer to the same physical process. Colossians 1:17 reminds us of God's ongoing work of upholding and maintaining the creation: "In [Christ] all things hold together." In order to maintain the everyday sense of saying "God creates each star and each tree," we will also use the word *create* to describe the ongoing processes by which new stars, planets, plants, and animals are formed.

THE UNIVERSE IS OLD

The dynamic changes and developments in the universe have been going on for a long time. In chapter 5 we described how geologists, over the past three centuries, have accumulated many kinds of evidence from rocks that the earth is billions of years old. In a similar fashion, over the past century astronomers have studied planets, stars, and galaxies and have found many strands of evidence that the universe is billions of years old. This consensus of astronomers is based on many independent measurements and has stood the test of time, a good indication that these results are reliable. In this section we'll describe some of this evidence for the great age of the universe.

Evidence from the Size of the Universe

We've already discussed the vastness of the universe earlier in this chapter. We noted that the most distant galaxies are over 10 billion light years away, indicating that the light left these galaxies over 10 billion years ago in order to reach us today. The straightforward interpretation of these data is that the universe must be at least 10 billion years old.

While some people have argued that perhaps these galaxies aren't really that far away, all of the methods used to measure distance agree that galaxies are billions, not thousands, of light years away. Others have argued that perhaps the light moved much faster when it first left these galaxies, so that it could reach us in much less time than 10 billion years. But this idea conflicts with other data that we have. As described in chapter 3, ample evidence supports the idea that physical processes such as quantum mechanics and electromagnetism function the same way in distant galaxies as they do on earth. Those physical processes depend on the speed of light and would look very different if the speed of light had changed. Instead, they look the same in distant galaxies as they do on earth, indicating that the speed of light has been constant over the history of the universe.

A Time Machine?

Because light takes time to travel, looking at distant stars and galaxies gives us a unique view into their past. We don't see them as they are *now* but as they were *then* when the light was emitted and began traveling toward us. This gives us a sort of "time machine," a way to look back in time to see what the universe was like millions and billions of years ago. By looking at the pasts of many galaxies, we can deduce how galaxies in general looked in the past. What we find is that galaxies were quite different billions of years ago; they tended to be smaller, bluer, and brighter, with lots of stars forming in them.

Evidence from the Moon and Planets

Studies of the Moon and planets also give evidence for great age. Geologists can use some of the same methods to measure the age of rocks on the Moon, Venus, and Mars as they use on Earth. That's because the asteroid collisions, volcanoes, and erosion they observe on Earth also occur on the Moon and planets. Photos taken by spacecraft while orbiting Mars show channels and gullies on the planet's surface. Similar channels on Earth are usually made by flowing water. Yet there is no liquid water on the surface of Mars right now.

What does this have to do with age? It is evidence that Mars was much different in the past than it is today. The atmosphere used to be much thicker and warmer, similar to Earth's, but now it is much colder and thinner. This dramatic change in planet-wide climate took millions or billions of years. Thus the rocks testify that the planet Mars must be at least this old.

In the last decade rovers have crawled the Martian surface and spacecraft have photographed it in sharp detail. Check out "Water on Mars?" on our website (www.faithaliveresources.org/origins) for what scientists are learning about the history of Mars.

Evidence from the Orbits of Asteroids

The orbits of asteroids also show evidence of a long history. When an asteroid is discovered, its path through the sky shows its orbit around the Sun. Once astronomers know the orbit of an asteroid they can calculate its orbit in the past and into the future to see whether it will hit the earth. By calculating the orbits backward, astronomers have found several asteroids that converged at the same location several million years ago. Apparently two larger asteroids collided at this spot and shattered into the smaller asteroids we see today. If God had created asteroids just a few thousand years ago, why would he have put them in orbits that suggest a collision several million years ago? The evidence clearly points to a long history for asteroids.

Evidence from Meteorites

Radiometric dating is used to study rocks on Earth as well as rocks from elsewhere in the solar system. (See ch. 5 and "Radiometric Dating" on our website, www.faithaliveresources.org/origins.) Studies have been done on the rocks that astronauts brought back from the Moon and on asteroids that have fallen to Earth. As with Earth rocks, scientists use multiple radioactive isotopes to cross-check age measurements. At least three different isotopes have been used to measure the age of Moon rocks, and at least five different radioactive isotopes have been used to measure the age of meteorites. The results all agree: the oldest Moon rocks and asteroids are 4.6 billion years old. This is our best measure of the age of the solar system as a whole. The universe itself must be at least this old.

Scientists call rocks *asteroids* when they are in orbit around the Sun, *meteors* when they are falling through Earth's atmosphere, and *meteorites* once they've landed on the ground.

Evidence from Star Clusters

Another important measure of age in the universe comes from star clusters. Because all stars in a star cluster form in the same nebula at about the same time, they all have about the same "birthday." But they don't all have the same lifespan. High-mass stars burn bright and fast like a "flash in the pan," while low-mass stars burn slowly and steadily. Consider how this will look in a star cluster. A cluster starts with many stars with the same birthday but of all different masses. Over time the high-mass stars die off first, leaving behind the low-mass stars. This means that if many high-mass stars are present, the cluster must be young because they haven't burned out yet. If most of the stars are low-mass, the cluster must be old. Careful studies of star clusters show that some clusters are younger and some are older, with the oldest ones having an age of about 12 billion years.

Multiple Lines of Evidence

The most distant galaxies, the planets and asteroids of our own solar system, and the oldest star clusters *all* are several billion years old. Astronomers have many different methods for measuring the age of various objects, and they all support ages of billions of years, not thousands. Even if the assumptions of one or two methods were faulty, it is highly unlikely that all of the methods would be affected. Like the geologists in the 1700s, astronomers today have found multiple lines of evidence against a young earth and young universe.

It may seem as though we are once again describing a conflict between science and theology. Scientific results that indicate great age do conflict with the Young-Earth Interpretation of Genesis 1 discussed in chapter 5. But remember that in chapters

5 and 6 we presented many other interpretations of Genesis 1; several of these are *not* in conflict with the great age found in the book of nature. In chapter 6 we also explained why we believe that the best biblical scholarship, quite independent of modern science, indicates that Genesis 1 was never meant to convey scientific information to the original audience. Its intent for the first listeners, and for us, is to teach the *who* and *why* of creation, not the *how* and *when*. Taken in this context, there is no conflict between Genesis 1 and the astronomical evidence for great age.

> The wide variety of evidence for a long history of the universe also challenges the Appearance of Age Interpretation of Genesis 1 discussed in chapter 5. See "A Detailed False History?" on our website (www.faithaliveresources.org/origins).

THE UNIVERSE HAD A BEGINNING

The universe is old, but it is not *infinitely* old. It had a beginning in time called the *Big Bang*. In this book the term *Big Bang* refers to a *scientific* model for the early history of the universe, not an atheistic worldview that the universe somehow created itself. The Big Bang model, if scientifically correct, is simply a description of *how* God governed the early universe. We'll discuss three major pieces of scientific evidence for the Big Bang model.

Evidence for Expansion

The first piece of evidence is that virtually all of the galaxies we see are moving away from our own galaxy in a specific pattern. The most distant galaxies are moving away the fastest, and nearby galaxies are moving away more slowly. This pattern of motion tells us that the universe is expanding; the fabric of space is expanding in all directions, and the galaxies are being pulled along with it. Here's an analogy: the universe is like a loaf of raisin bread rising. The dough is like the fabric of space, and the raisins are like the galaxies. Just as the dough rises, carrying all of the raisins apart from one another, so the universe expands, causing all of the galaxies to move away from us. A raisin on the far side of the loaf will move a few inches while the dough rises, but a nearby raisin will move less than an inch in the same rise time. Thus the distant raisin moves away faster than the nearby one. In the same way distant galaxies move away faster than nearby ones, indicating that the whole fabric of space is expanding.

Are we at the center of the universe? All galaxies in the universe are moving away from Earth. This gives the impression that our galaxy is located at the unique center of everything, but the pattern of the expansion tells us otherwise. In the analogy of raisin bread, every raisin moves farther from all of its neighbors. Each raisin, not just the one at the center of the dough, sees all of the other raisins moving away from it. The same is true for the universe: every galaxy sees all of the other galaxies moving away. The expansion we see does not mean that our galaxy or any other galaxy is at a unique center of the universe.

What happens if we mentally rewind the universe back in time, using the equations and laws of physics to reverse the expansion? We would see the fabric of space contracting rather than expanding. The galaxies and stars would be packed closer together, eventually so packed together that they would not even be separate entities. The universe would just be hot gas. If we rewind the motion farther backward in time, that gas would be even more densely packed and hotter yet. This hot, dense beginning, followed by expansion, is what astronomers call the Big Bang.

This description of a hot, dense beginning of the universe is an extrapolation of the current expansion of the universe backward in time. Is this really what happened? Maybe the universe started more recently, with stars and galaxies in place, and the expansion continued from there. Do astronomers have any other evidence that the expansion can be traced all the way back to such a hot, dense beginning? They do.

Evidence for a Hot Beginning

In the Big Bang model a hot early universe would have given off heat radiation just as a toaster element radiates heat. That heat radiation would still be around in the universe today, spread out everywhere in space. It would have cooled off by now, just as a hot gas cools off as it expands, but it should still be detectable. In contrast, if the Big Bang model is wrong and the universe started more recently, with stars and galaxies in place and the expansion continuing from there, then heat radiation should not be everywhere in space. Based on this, the Big Bang model made a *prediction* that telescopes would discover this heat radiation. This prediction was confirmed in 1965 with the discovery of the Cosmic Microwave Background Radiation (CMBR). As predicted, radiation is coming from all over the sky, not from any particular star or galaxy but from the universe itself. As predicted, the CMBR has the spectral signature of heat radiation. And, as predicted, the expansion of the universe has cooled the radiation tremendously, down to a temperature of only 2.726 degrees above absolute zero. (That's minus 270 degrees Celsius, or minus 455 degrees Fahrenheit!) This heat radiation is the second major piece of evidence for the Big Bang model.

Evidence for Fusion at the Beginning

The third major piece of evidence is the amount of helium in the universe. The ordinary matter in the universe is about 75 percent hydrogen, 24 percent helium, and 1 percent other elements. Why this percentage and not some other ratio? We've already talked about how helium is formed in stars by fusion of hydrogen. But even in a universe billions of years old, the fusion in stars happens much too slowly to account for this much helium. How else could helium have been produced? Using the Big Bang model, astrophysicists calculate that the conditions of the universe about three minutes after the Big Bang were very similar to the interior of a star and just right for fusion reactions. The temperature and density of the hydrogen gas allowed it to fuse into helium and into trace amounts of deuterium and lithium. The calculations of the Big Bang model even make precise predictions for the relative percentages of helium, deuterium, and lithium that would be produced. The model predicts that about 24 percent of the gas would be helium, in agreement with what astronomers observe. The model predicts less than 1 percent of deuterium and lithium, and this also matches astronomers' observations. This impressive agreement between model and observations is compelling evidence that the universe really was hot and dense enough for fusion in early times.

Age of the Universe as a Whole

The Big Bang model gives us one more measure of age, the age of the universe as a whole rather than the age of the individual planets or stars or galaxies in it. By tracing the expansion of space backward in time and taking into account the slight changes in the expansion rate that have been measured, astronomers calculate that all of the matter in the universe must have been compressed together 13.7 billion years ago. That is the best current value we have for the age of the whole universe. This measurement is consistent with the ages of the solar system, stars, and galaxies described earlier in this chapter. The universe is older than the objects in it, but not many times older.

This is also evidence that the universe is not *infinitely* old. The Big Bang model affirms the Genesis account that our universe had a beginning. Interestingly, many astronomers in the mid-1900s advocated a different model called the *Steady State Universe*. In this model the universe was thought to be infinite, unchanging, and not having had a beginning. The Steady State model was developed at a time when scientists had discovered some evidence for the Big Bang, namely the expansion of the universe. But evidence for the Big Bang was not yet overwhelming because the heat radiation had not yet been observed. In some ways the situation in the mid-1900s was similar to Galileo's day. Multiple models could explain the existing data, and new evidence was just starting to arrive in favor of one model over the others. When the scientific data cannot yet determine which model is best, considerations from *outside* the field of science, such as worldviews and tradition, play more of a role than when the scientific evidence is solid.

In the 1940s and 1950s the atheistic worldviews of some astronomers led them to favor the Steady State model over the Big Bang model. They disliked the idea of a beginning to the universe and wanted a scientific explanation for what happened before the bang. Some Christians at that time were more open to the Big Bang model. One of the earliest of these was Georges Lemaître, a Belgian cleric who developed some of the first mathematical models for the expansion of the universe. After the discovery of the cosmic microwave background radiation in 1965, most astronomers of all worldviews became convinced of the Big Bang. Although the atheistic and agnostic worldviews of some prominent astronomers made them prefer the Steady State model for philosophical reasons, the increasing *scientific* evidence convinced them that the Big Bang model was better. The scientific data led them to a truer picture of God's universe, a universe that had a beginning.

Does the scientific evidence for a beginning *prove* the existence of God? For a discussion of this question, visit our website (www.faithaliveresources.org/origins). Look for the article "Does the Big Bang Prove That God Exists?"

Scientific Evidence in a Christian Package

Sometimes when the Big Bang model and other results from astronomy are presented in the media or in popular books, they are given an atheistic spin. (Recall Carl Sagan's description of the vastness of the universe earlier in this chapter.) This can make it difficult for Christians to separate legitimate scientific evidence from the atheistic worldview in which it has been packaged. If the science and the worldview can't be separated, it certainly is safer to toss out the science than to agree with the atheistic worldview conclusion. But this is like throwing the baby out with the bathwater.

More often, astronomy is explained in neutral language that is neither atheistic nor theistic. Scientists simply say "stars formed" instead of either "God formed the stars" or "Stars formed without a guiding hand or purpose." But if you've heard it once in atheistic language, then hearing it again even in neutral language can still sound God-denying. In this book, we are presenting the scientific evidence in a Christian package.

> To compare how the same scientific model can be written in three ways—first with neutral language, then with an atheistic spin, and then with a theistic spin—look for "A Brief History of the Universe Spun Three Ways" on our website (www.faithaliveresources.org/origins).

THE UNIVERSE IS FINELY TUNED FOR LIFE

When astronomers consider the universe, they can imagine all sorts of ways it could be different. They consider what would happen if the force of gravity were stronger or weaker than it actually is, or what would happen if atoms weighed more or less than they actually do. When they calculate what would happen in such imaginary universes, astronomers usually find that these universes would be so bizarre that they would not allow human beings, or any other life, to develop and survive. By contrast, our actual universe is amazingly well suited for life.

What properties does the universe need to support life? At a minimum, "life as we know it" requires two things:

▶ a stable energy source, such as the light from a nearby long-lived star

▶ a variety of atoms that can combine into many varieties of molecules to allow complex chemistry

It might be possible for some bizarre life-form to develop without an energy source or made of only one kind of atom, but it's hard to imagine what that life would look like. Stars are important for meeting both of these conditions for life: stars provide energy and produce all the different types of atoms through fusion and fission. Astronomers have found that the properties of our universe appear to be *fine-tuned* for the development of stars and a variety of atoms. The particles, the forces that hold them together, and the overall parameters of the universe are set up in just the right way for life to exist. Sometimes this is called a "Goldilocks universe": not too hot, not too cold, but just right. We'll look at five factors that make the universe just right.

Expansion Rate of the Universe

Imagine a universe with a different expansion rate. Recall that astronomers have detected the expansion of the universe; space itself is expanding and pulling all galaxies farther apart from each other. If the universe had expanded a bit more rapidly after the Big Bang, it would have pulled the gas clouds apart before they could form into stars and galaxies. If it had expanded a bit more slowly, the universe would have been overwhelmed by its own gravity and collapsed into a "Big Crunch" before stars had a chance to form. Instead, the actual universe expanded at just the right rate to allow stars to form.

Force of Gravity

Imagine a universe in which gravity pulled with a different strength. A star is a balancing act between gravity pulling in and gas pressure pushing out. If the gravitational force were a bit weaker, it wouldn't be able to hold a star together; the gas pressure in a star would blow it apart. If the gravitational force were a bit stronger, it would easily hold stars together, but the stars would be denser and burn faster. This would cause them to burn out quickly. The force of gravity we observe is set just right to allow stable, long-lived stars.

Fundamental Physical Forces

Imagine a universe where the physical forces have different strengths. Three physical forces are involved in fusion reactions: the strong nuclear force, the weak nuclear force, and the electromagnetic force. In the actual universe the strengths of these forces are balanced so that protons and neutrons can successfully fuse together to make heavier atoms and atoms can exist stably without falling apart. If the strengths of these forces were a bit different, fusion reactions would not occur properly, and the only stable element in the universe would be hydrogen. If the strengths weren't just right, a variety of atoms would not exist to support complex chemistry.

Nuclear Reaction Rates

Imagine a universe without carbon. Why care about carbon? It is one of the few elements that can form long-chain molecules with hydrogen (hydrocarbons). Such long-chain molecules are very well suited for complex chemical life, and hydrocarbon chains form the backbone of the molecules in all life on earth. Carbon atoms are made in the cores of stars, as part of a sequence of fusion reactions involving other elements. First helium atoms fuse together to make carbon, and then carbon fuses with helium to produce oxygen. In order for stars to manufacture a lot of carbon, the fusion reaction rates must be set just right so that a lot of helium fuses into carbon *and* the carbon is not all used up making oxygen. Scientists have discovered that carbon has a nuclear excited state that is just right for this to happen. This excited state allows lots of carbon to be produced. On the other hand, oxygen does not have such an excited state, so only some of the carbon is made into oxygen. The nuclear properties of these atoms are tuned just right to make a lot of carbon, an element fundamental to life.

Water Molecule

Imagine life without water. Water is used by all life-forms on earth. It has properties that make it extremely good at dissolving and transporting many kinds of molecules and ions. Water also happens to be transparent to sunlight in the wavelengths our eyes can see, but it is *not* transparent to other wavelengths of light, such as infrared light, x-rays, or ultraviolet rays. This means that water in the atmosphere allows some sunlight to reach the earth's surface—the light with just the right wavelengths to stimulate important chemical reactions like photosynthesis—but blocks most of the ultraviolet and x-ray light, wavelengths that destroy big molecules. Water is transparent to exactly the sort of light that is important for life, and this is the same sort of light produced by long-lived stars like the Sun. Why is this? Based solely on the laws of physics, it's hard to think of a *physical* reason that the light emitted by stars should have any correlation with the transparency of a molecule so useful for life, and yet that is exactly what we have. One or both is fine-tuned for life as we know it.

Alternate approaches to fine tuning include the *anthropic principle* and the *multi-verse hypothesis*. Look for a discussion of these ideas in "Does the Scientific Evidence of Fine Tuning Prove the Existence of God?" on our website (www.faithaliveresources.org/origins).

Scientists have discovered even more fine-tuned parameters than the five we've discussed here. The careful construction of the universe is consistent with the biblical belief that God planned the universe to include intelligent human beings who can in turn relate to him. The universe itself testifies to God's amazing craftsmanship: God made a system that is deceptively simple and yet amazingly productive. It takes only a few numbers and equations on a single sheet of paper to write down the fundamental properties of the universe, the fundamental laws of physics, and the list of elementary particles (electrons, quarks, and so on). This simple but amazing system allows stars and molecules not merely to exist but to *assemble naturally over time*.

In the Big Bang all matter in the universe was in the form of simple elementary particles. But as the universe developed over time these particles combined to produce an abundance of stars and galaxies, a wide variety of atoms and molecules, planets with land and ocean and atmosphere—all the building blocks necessary for life to exist and a suitable planet to be its home. The physical forces and the properties of atoms are the same wherever and whenever we look in the observable universe; God didn't need to tweak the system as he went along. Instead, God designed it from the beginning to produce what we see today. Although God could have made each atom and each molecule, each star and planet, by a separate supernatural miracle, the universe testifies that he instead chose to work through a beautiful system of regular natural processes that we have the pleasure of studying scientifically. Our study of astronomy reveals evidence of God's design not because the universe is too hard for science to understand but because we *can* understand it.

QUESTIONS FOR REFLECTION AND DISCUSSION

1. What ideas in this chapter were new to you? What did you learn about the universe?
2. What helped you understand God better? Which of God's attributes became more vivid to you?
3. What examples of *atheistic language* have you heard used when talking about science? Can you think of particular authors, magazines, or TV shows that use science to attack religion?
4. Does the scientific picture of the vast cosmos make you feel more significant or less significant? Why? What is the basis for our significance as humans?
5. In chapters 5 and 6 we discussed scientific evidence for the great age of the earth, principles of biblical interpretation, and several different interpretations of Genesis 1. How do you think these relate to the Big Bang model?

ADDITIONAL RESOURCES

More on the place of humanity in the cosmos from a Christian perspective:

Danielson, Dennis. "Copernicus and the Tale of the Pale Blue Dot" (search on www.google.com).

More on astronomical evidence for the age of the universe from a Christian perspective:

Van Till, Howard. "The Scientific Investigation of Natural History," *Portraits of Creation.* Grand Rapids, Mich.: Wm. B. Eerdmans, 1990.

More on astronomy and cosmology from a mostly religiously neutral perspective:

American Astronomical Society. "An Ancient Universe: How Astronomers Know the Vast Scale of Cosmic Time," 2004. A readable 20-page overview of evidence for age.

Coles, Peter. *Cosmology: A Very Short Introduction.* Oxford: Oxford University Press, 2001. 130 pages.

Ferreira, Pedro. *The State of the Universe: A Primer in Modern Cosmology.* London: Cassell, 2006. 300 pages.

More on fine-tuning from a Christian or theist perspective:

Craig, William Lane. *The Teleological Argument and the Anthropic Principle.* www.leaderu.com/offices/billcraig/docs/teleo.html.

Leslie, John. *Universes.* London: Routledge, 1996.

Ross, Hugh. *More Than A Theory: Revealing a Testable Model for Creation.* Grand Rapids, Mich.: Baker Books, 2009

More on fine tuning and the multi-verse hypothesis, from a non-Christian perspective:

Green, Brian. *The Hidden Reality: Parallel Universes and the Deep Laws of the Cosmos.* Knopf, 2011.

Rees, Martin. *Just Six Numbers: The Deep Forces that Shape the Universe.* New York: Basic Books, 2000.

CHAPTER 8

COMPETING VIEWS ON EVOLUTION

Jennifer grew up in a Christian home. During her teenage years she made a personal commitment to Jesus. Her family, pastor, and Sunday school teachers encouraged her to enroll in a nearby university when she finished high school. But they warned her about atheists at the university, including some on the faculty, who would attack her faith. She was told that atheists would use evolution to try to convince her that God doesn't exist. Her youth group leader showed a video that defended creationism and argued that evolution couldn't happen.

Jennifer and the other characters in this story are fictional and based on the experiences of several people.

During her first semester at the university Jennifer joined a Christian student fellowship group. There she met Professor Bensen, a faculty mentor to the group. Professor Bensen clearly had a vibrant faith and cared deeply for the students. He encouraged them to keep their spiritual life strong through regular Bible study, prayer, and worship.

Professor Bensen, a successful scientist studying disease-causing bacteria, had a reputation as a good teacher. In part because of his inspiration, Jennifer decided to become a doctor. She registered for a biology course taught by Professor Bensen, thinking that it would be both interesting and safe. On the first day of class Jennifer noticed that the textbook assumed that the theory of evolution was true. She guessed that Professor Bensen simply had to use a textbook approved by his department and that he would deal with the issue somehow. But as the semester wore on, Professor Bensen said things in class suggesting that he actually believed the theory of evolution. Finally, Jennifer went to his office to ask him about it.

Professor Bensen carefully explained that a great deal of scientific evidence clearly supports evolution. He said that it is right for a Christian to believe that the theory of evolution is correct because scientific evidence supports it and because the Bible doesn't teach against it. When Jennifer left the professor's office, her head was buzzing. She was more confused than when she had entered. This was the first time she had heard anything like this.

Jennifer had great respect for Professor Bensen. She had seen the professor's faith in action many times. But if he was right, then her pastor, Sunday school teachers, and parents must be wrong, and she had great respect for them too. She didn't know whom to believe or where to go for answers.

Perhaps you identify with Jennifer. In this chapter we'll attempt to provide some answers by

▶ defining five different meanings of evolution.
▶ taking a closer look at evolutionism.
▶ describing positions Christians hold on evolution.
▶ discussing theological issues related to progressive and evolutionary creationism.

In the next chapter we'll summarize scientific evidence regarding plant and animal evolution.

EVOLUTION AND SCIENCE, EVOLUTION AND RELIGION

Ever since Charles Darwin published *On the Origin of Species* in 1859, the theory of evolution has been a battleground for competing religious and philosophical claims. Those battles have often spilled over into political and legal arenas.

Many people, including scientists, say that religious or political debates about evolution are unnecessary. They say that evolution is a scientific model just like any other. Just as the scientific theory that gravity causes the earth to orbit the sun has nothing to do with religion, and the scientific theory that evaporation and condensation form rain has nothing to do with religion, so also the theory of evolution has nothing to do with religion. It says nothing about God one way or the other. Like any other scientific theory, it should be accepted or rejected on the basis of the data. If religion and politics stayed out of it, no debate would occur.

But for many people, Christians and non-Christians alike, evolution and religion are linked. These people say that evolution is different from most other scientific theories because it contradicts particular Christian beliefs. Some atheists claim that *if* evolution is true, then Christianity and any belief in God must be false. They go on to reason that because evolution *is* true, then Christianity must be false. Some Christians agree with the first half of this claim. They agree that *if* evolution is true, then Christianity must be false, but they respond by saying that because Christianity is true, then *evolution* must be false.

As we pointed out in the introduction to this book, there are more than just these two options to consider. Many Christians respond in other ways to evolution. The range of Christian responses is often divided into three broad categories:

▶ young-earth creationism
▶ progressive creationism
▶ evolutionary creationism

Before we look at each of these, we should be clear about what the word *evolution* means.

> In this chapter and the next we will focus specifically on the issue of plant and animal evolution. We'll save the topic of human origins for chapters 11 and 12.

Evolution—What Does It Mean?

If someone says either "I don't believe in evolution" or "Evolution has been proven scientifically," it's important to find out what he or she means by the term *evolution*. One reason debates about evolution can be confusing is that people frequently mix different definitions without even realizing they are doing so. To avoid such confusion we distinguish five different definitions of evolution in this chapter:

▶ **Microevolution.** Small changes in species, caused by the mechanisms of evolution, accumulate over a few decades or centuries, allowing species to adapt to an environment and sometimes to split into two or more species.

▶ **Pattern of Change Over Time.** The fossil record shows that species go through changes over long periods of time, with some species becoming extinct and new species appearing from time to time.

▶ **Common Ancestry.** All living and extinct species are linked in a "family tree"; modern species descended from earlier species and all species descended from a common ancestor.

▶ **Theory of Evolution.** A scientific model asserting that the mechanisms of evolution, operating over the long history of life on earth, explain common ancestry and the pattern of change over time.

▶ **Evolutionism.** Attempts to use the theory of evolution to support atheistic claims that there is no Creator and no purpose to human existence.

By understanding these different definitions you will be much better equipped to understand the debates over evolution. Let's consider each in more detail.

Microevolution

Microevolution is based on two fundamental mechanisms:
▶ differential reproductive success
▶ random mutation

Differential reproductive success means that some plants or animals are better suited to their environment than others and that the ones that are better suited to their environment will tend, on average, to produce more offspring. This is sometimes called "natural selection" (or "survival of the fittest," although that term is not as accurate). Organisms that are less suited to their environment can still survive, thrive, and reproduce, but organisms that are better suited will tend to survive longer, attract mates more successfully, and produce more offspring.

Random mutation refers to one way in which the genetic makeup of plants and animals can change. Every plant and animal has a collection of *genes* in its cells. These genes are segments of a special molecule called *DNA* that serves as a sort of instruction manual to tell each cell how to manufacture the chemicals it needs. If we think of an organism as a building, then the genes are the blueprint for the building. Each gene is like one page of the blueprint describing some particular details of the organism. In plants and animals with two parents, each offspring gets half of its genes from each parent. Usually each gene in the offspring is identical to a gene in one of the parents. When a *mutation* happens, a gene from the parent is changed in the offspring. Some mutations are harmful to the offspring and others beneficial, but the great majority of mutations are neutral—neither harmful nor beneficial. For example, a mutation in a gene might make a slight change that has no measurable effect. Or a mutation might add an extra copy of a gene; both the original gene and the copy function normally until one or the other mutates again. Saying that the mutations are *random* means at least two things. First, scientists cannot predict when a mutation will happen or what form it will take. Second, the actions of the parents do not determine what sort of mutation the offspring will have. For example, a parent animal that spends a lot of time in the dark is not more likely to have offspring with a mutation for better nighttime vision. If the offspring has a vision mutation, the mutation might make the night vision better, worse, or leave it unchanged.

Microevolution refers to the process by which these two mechanisms cause species to change and adapt to their environment over time. Microevolution happens fastest in species that reproduce new generations quickly. Some of the most notable examples of microevolution include disease-causing bacteria that have developed a resistance to antibiotics. When a large number of bacteria are given a weak dose of antibiotic, a few will have a mutation that makes them resistant to the antibiotic, and these few will best survive and reproduce. When they once again grow to large numbers, a higher dose of antibiotic is required to treat the infection. Again, a few bacteria will have another mutation that makes them even more resistant, and these few will best survive and reproduce. After several generations the surviving bacteria can be completely resistant to the antibiotic. Over the last few decades several types of infectious bacteria have developed resistance to many different antibiotics. In a similar fashion some weeds and insects have developed resistance to certain herbicides and pesticides.

Many times within human history scientists have observed microevolution as it occurred. For example, house sparrows were introduced to North America in 1852 and have spread over large regions since then. Today sparrows in the colder north have larger bodies on average than sparrows in the warmer south.

For a frequently cited example of microevolution see "The Microevolution of Peppered Moths" on our website (www.faithaliveresources.org/origins). Or visit www.millerandlevine.com/km/evol/Moths/moths.html.

Sometimes microevolution can lead to one species splitting into several species. About 3,700 years ago a bay of Lake Victoria, one of the Great Lakes of Africa, was closed by a sand bar, forming Lake Nagubago. When this happened some cichlid fish were trapped in Lake Nagubago and isolated from their parent stock in Lake Victoria. They have since microevolved into five new species that differ in their coloration and mating rituals and do not inter-breed with each other.

When some people look at evolution, they see an ugly system of competition and death. But others see a beautiful system of adaptation and intricacy. Learn more by reading "Is Evolution Ugly or Beautiful?" at our website (www.faithaliveresources.org/origins).

Pattern of Change Over Time

The phrase "*pattern of change over time*" refers to changes seen in species in the long history of life, going back billions of years. The fossil record shows that modern species look somewhat like species in the recent past, less like species in the more distant past, and only a little bit like species in the very distant past. Species in the distant past are generally simpler than modern species. Whenever a new species first appears in the fossil record, it looks a lot like other species that existed at the same time, but then that species progressively changes over time, and sometimes the species dies out. This pattern of change over time says that species have changed over the history of the earth, but it makes no conclusion about the *reason* or the *mechanism* for that pattern of change. It describes the observations, but it is not a scientific model that *explains* these observations.

Common Ancestry

Common ancestry refers to the idea that all past and present living organisms descended from a common ancestor; that is, all species are linked into a sort of family tree. Modern species of dogs and wolves and coyotes are descended from some ancestral, wolf-like species that no longer exists. Modern species of lions, tigers, and house cats are descended from some ancestral cat-like species that no longer exists. All dogs, cats, and other mammals are descended from a common ancestor even longer ago. All mammals, birds, reptiles, and fish are descended from a common ancestor that lived still longer ago. Common ancestry, if true, would partially explain the pattern of change over time seen in the fossil record. It would explain, for example, why a new species that appears in the fossil record looks a lot like other species of that time. But even if common ancestry fits the data, it is an incomplete model. By itself, it doesn't give a mechanism to explain *how* species change and split over time.

Theory of Evolution

The *theory of evolution*, as most scientists use the term, states that random mutations and differential reproductive success not only produce small changes in species in just a few centuries (microevolution) but also produce large changes in species over millions of years. All living and extinct species share common ancestry; the pattern of change and the development of new species was produced by the mechanisms of evolution operating over millions of years.

When Darwin proposed his theory of evolution, he wrote a great deal about natural selection (differential reproductive success), but he did not speculate very much about the causes of random mutation. In Darwin's time little was known about genes, and the DNA molecule wasn't discovered until many decades later. When scientists today write about the theory of evolution, sometimes they are referring to Darwin's original theory, but most of the time they are referring to the modern form of the theory. The modern form of the theory combines Darwin's original theory with what has been learned since then about genes, DNA, mutations, the fossil record, and differential reproductive success.

Evolutionism

The word *evolutionism* is in a different category from the others. It does not refer to science but to a set of worldview beliefs. It refers to the ways in which some people try to use the theory of evolution to support certain atheistic beliefs. Among the claims of *evolutionism* are the following:

▶ There is no Creator who cares for the world.

▶ Humans arose by a purely natural process, without guidance or governance from God.

▶ Human existence has no higher purpose.

▶ Human morality is merely the result of heredity and environmental influences; there is no absolute morality.

These are not scientific but philosophical and religious statements. When the theory of evolution is used to argue for atheistic beliefs, it is rightly called *evolutionism*.

A CLOSER LOOK AT EVOLUTIONISM

Many examples of evolutionism appear in popular books and in media stories on evolution. In his essay "In Praise of Charles Darwin," biologist Steven Jay Gould wrote that the theory of evolution seems to imply that God is not involved:

No intervening spirit watches lovingly over the affairs of nature. . . . No vital forces propel evolutionary change. And whatever we think of God, his existence is not manifest in the products of nature.
—Stephen Jay Gould, "In Praise of Charles Darwin,"
Discover Magazine, February 1982.

Another common example of evolutionism is found in arguments that confuse two different definitions of *chance*. In 1972, for example, the biochemist Jacques Monod wrote that

Chance alone is at the source of every innovation, of all creation in the biosphere. Pure chance, absolutely free but blind, is at the very root of the stupendous edifice of evolution . . . [These biological discoveries] make it impossible to accept any system . . . that assumes a master plan of creation.
—Jacques Monod, *Chance and Necessity,* 1972.

On the one hand Monod seems to be referring to the chance or random nature of biological mutations, certainly essential to evolution. In that sense "random" means simply that mutations are *unpredictable*; no scientist can predict when a mutation will happen or what form it will take. On the other hand Monod is talking about the purpose and plan of creation, something much bigger. In this sense he is using "chance" to refer to a *lack of meaning or purpose*.

In chapter 2 we talked about the error of mixing these two definitions of *chance*. Monod's argument mistakes the *scientific* unpredictability of mutations for the *worldview belief* that mutations (and thus evolution) are "blind" and without purpose. But Christians can look at the same scientific unpredictability and come to a completely different worldview conclusion. The Bible teaches that God governs events that are unpredictable by humans, from the casting of lots to the arrival of thunderstorms. Christians see mutations as something governed by God and can see the mechanisms of evolution as part of God's design and ultimate purpose for species. The biological theory of evolution does *not* rule out a master plan for creation.

For another example of evolutionism in the media, see the story "Darwin and Floating Plant Seeds" on our website (www.faithaliveresources.org/origins).

To be fair, many agnostics and atheists agree that the theory of evolution does not support evolution*ism*. A number of scientists who do not believe in God have said quite clearly that the theory of evolution is compatible with a religious belief in God.

However, other scientists and philosophers do mix together the theory of evolution with the philosophical and religious claims of evolutionism. This heats up the religious debate over evolution. People who may not understand the subtleties of the theory of evolution—but who are certain that they disagree with the atheistic assertions of evolutionism—ask: Have scientists really proven these claims? Does science really teach that there is no God? Are they teaching that to our children in science classes at school?

WHERE CHRISTIANS AGREE AND DISAGREE ABOUT EVOLUTION

As we noted earlier in this chapter, the range of Christian responses is often divided into three broad categories. Below we list how these three groups typically respond to specific definitions of evolution. In the Appendix you'll find a detailed list of multiple positions within each group.

Young-earth creationists
▶ accept microevolution.
▶ say that the earth is young.
▶ reject that the fossil record shows a pattern of change over time.
▶ reject common ancestry.
▶ reject the theory of evolution.
▶ reject evolutionism.

Progressive creationists
▶ accept microevolution.
▶ say that the earth is old.
▶ accept that the fossil record shows a pattern of change over time.
▶ are split about common ancestry (some accept and others reject it).
▶ reject the theory of evolution as a complete model for biological history, saying that while some evolution did happen, God must have miraculously guided or intervened at various points.
▶ reject evolutionism.

Evolutionary creationists

▶ accept microevolution.

▶ say that the earth is old.

▶ accept that the fossil record shows a pattern of change over time.

▶ accept common ancestry.

▶ accept the theory of evolution as a scientific model.

▶ reject evolutionism.

All Christians are united against *evolutionism* but disagree with each other about the best strategy for combating it. To illustrate this, let's take a simplified argument for evolutionism (with construction similar to an argument used in ch. 4 about the motion of the earth).

Premise 1: If the theory of evolution is true, then Christianity is false when it says that God created all of the plants and animals.

Premise 2: Science shows that the theory of evolution is true.

Conclusion: Christianity is false.

Young-earth creationists and progressive creationists combat evolutionism by attacking the second premise. They argue that the scientific evidence does not support the theory of evolution. Evolutionary creationists combat evolutionism by attacking the first premise. They argue that the Bible does not teach against evolution and that God could work through biological evolution just as he works through other scientifically understandable natural processes.

In the next chapter we will discuss what the scientific evidence has to say about common ancestry and the theory of evolution. For now we'll focus on theological issues.

SOME THEOLOGICAL ISSUES WITH PROGRESSIVE AND EVOLUTIONARY CREATIONISM

In chapter 5 we discussed *young-earth creationism* (p. 118). For the rest of this chapter and the next we will concentrate on *old-earth creationist* views held by progressive and evolutionary creationists.

Interpretation of Genesis 1

How to interpret Genesis 1 is one theological issue. Many progressive creationists argue for *concordist* interpretations of Genesis 1 (see ch. 5). Most evolutionary creationists and some progressive creationists say that the principles of biblical interpretation show us that *non-concordist* interpretations of Genesis 1 (see ch. 6) are actually better.

The Role of Miracles

Another theological issue is the role of miracles in God's governance of the world. The Bible tells us that throughout human history God sometimes uses miracles, such as when he rescued the children of Israel from Egypt, while at other times God works through seemingly ordinary events, such as in the book of Esther. For that reason many progressive creationists argue that we should expect that God created modern life-forms through a combination of ordinary, natural processes *and* miracles. Evolutionary creationists, on the other hand, argue that while this is true of *salvation* history, it doesn't necessarily follow that this is how God works in *natural* history. While God can certainly do miracles, nature shows that he usually chooses to govern the natural world through natural processes and did so for most of natural history. Perhaps God chose to reserve miracles for his personal interactions with humans in salvation history.

Some progressive creationists argue that there ought to be evidence for God's miraculous creative action in nature. This will serve as evidence that creation tells about God's glory (Ps. 19:1;

Rom. 1:20) and as evidence against atheism. Some evolutionary creationists counter that there ought *not* to be evidence for miracles in the natural world, because this would be proof of God's existence. Having proof goes against the idea of having *faith*. Neither of these arguments is especially convincing. God could, if he chose, provide miraculous and unambiguously clear proof of his existence to *every* human being, but he chooses not to. On the other hand, the Bible tells us that God sometimes provides miraculous evidence of his existence and power, as for example in the lives of Moses, Elijah, and Jesus.

As we discussed in chapter 2, simply providing a scientific explanation for something does not exclude God from the process. We can provide scientific explanations for what causes the rain and for why planets orbit the sun, but God is every bit as much in charge of those natural events as he is in charge of miracles.

Theological Dangers

Purely theological arguments don't strongly favor either progressive creation or evolutionary creation. The biggest theological danger faced by the idea of progressive creation is that it results in a sort of "god of the gaps." It tends to look for God only in those events that are not scientifically explainable and seems to concede to atheists anything that is scientifically explainable. The biggest theological danger faced by evolutionary creation is that it becomes too much like *deism*: a belief that God started the universe and the laws of nature and then let it run on its own after that.

The way to avoid both of these theological dangers is the same: a solid biblical understanding that God is in charge of both natural events and miracles. If Christians hold on to that truth, progressive creationists and evolutionary creationists alike will see God's power and creativity in nature, no matter what methods God used. We can praise God if he chose to create using miracles, and we can praise God if he chose to create using the mechanisms of evolution. This frees us to examine the scientific evidence without fear and let God's book of nature teach us what it has to say.

THE UNITY OF BELIEVERS

What about Jennifer? In the story at the beginning of this chapter, Jennifer's Sunday school teachers actually have a lot in common with Jennifer's professor. They agree about who created everything, who redeemed them, and how they should live out the Christian life. They also agree that the atheistic philosophy of *evolutionism* is wrong, but they disagree how best to challenge it. Jennifer's Sunday school teachers believe that it's best to confront the theory of evolution. Professor Bensen believes that the theory of evolution is a good scientific model and instead confronts the philosophical claims of evolutionism directly. By maintaining a charitable attitude toward each other, Christians who advocate different responses to evolution need not break their unity as believers who work side-by-side to advance God's kingdom.

Imagine what might have happened if Jennifer hadn't met Professor Bensen. She might have taken a course from a stridently atheistic professor who promoted evolutionism. If so, she might have dropped the course and given up the idea of becoming a doctor. More likely, she might have taken a course from a professor who simply presented the scientific evidence for evolution and never mentioned religion. As the evidence piled up, it could have caused Jennifer to question *everything* she had learned from her church back home. Neither outcome is desirable.

Jennifer's parents and teachers were rightly concerned about evolutionism, but they put Jennifer in a painful position by giving her only two options: young-earth creationism or atheistic evolutionism. When students are forced to choose between these two, they may either turn away from a career in science or pursue science but turn away from God. A far better approach is to teach young people about a *range* of Christian positions on evolution, giving them some options for how to keep their faith when they encounter the theory of evolution.

QUESTIONS FOR REFLECTION AND DISCUSSION

1. Before you read this chapter, how would you have defined *evolution*? Would you have used just one definition or several?
2. This chapter described the experience of a fictional student named Jennifer. Do you know anyone who went through a similar experience in real life? What happened?
3. Have you heard or read examples of evolution*ism*? How would you respond to them?
4. Some Christians might say that God wouldn't use random mutations because such a system would be messy and disorderly. Others might say that God could select the outcome of every mutation and still others that it's amazingly beautiful that God crafted a system in which random mutations lead to well-adapted plants and animals. What do you think?
5. Should we expect scientific proof in nature that God is the Creator? Why or why not?

ADDITIONAL RESOURCES

Dembski, William and Michael Ruse, eds. *Debating Design: From Darwin to DNA.* Cambridge: Cambridge University Press, 2007. An anthology of different Christian views on evolution and intelligent design.

Keller, Tim. *Creation, Evolution, and Christian Laypeople.* BioLogos, 2009, www.biologos.org/uploads/projects/Keller_white_paper.pdf. Advice for pastors on how to answer the most pressing questions asked by laypeople about evolution.

Moreland, J. P., and John Mark Reynolds, eds. *Three Views on Creation and Evolution.* Grand Rapids, Mich.: Zondervan, 1999.

Ratzsch, Del. *The Battle of Beginnings: Why Neither Side Is Winning the Creation-Evolution Debate.* IVP Academic, 1996.

CHAPTER 9

EVIDENCE FOR PLANT AND ANIMAL EVOLUTION

onversations about evolution often degenerate into rancorous argument, slanted news coverage, and personal attacks. Much of this heat comes from confusion over the meanings of the word *evolution*, particularly when the scientific theory of evolution is equated with the worldview of evolution*ism,* as discussed in the previous chapter.

Yet when all of the heated language is cleared away, purely scientific questions still arise about evolution. In chapters 5 and 6 we discussed the interpretation of God's *Word,* and in chapter 8 we looked at worldview influences on evolution and the *theological* issues surrounding it. In this chapter we'll look at the *scientific* evidence coming from the study of God's *world.* What do scientists see in the natural world that leads them to the theory of evolution as a good model? We'll discuss

▶ scientific evidence from Darwin's time, including fossils, comparative anatomy, and biogeography.

▶ modern scientific evidence, including genetic similarity across species and genetic diversity within species.

▶ the question of common ancestry versus common function.

We'll close with a discussion of how this scientific evidence is viewed by progressive creationists and evolutionary creationists.

DARWIN-ERA SCIENTIFIC EVIDENCE FOR THE THEORY OF EVOLUTION

Darwin saw several lines of evidence that led him to develop the theory of evolution. We'll summarize three of them.

Fossil Evidence

Most fossils are formed when plants or animals are buried in muddy or sandy sediment. As more and more sediment is added, the pressure increases, water is driven out, and the sediment mineralizes and becomes rock. An impression of the plant or animal is preserved in the rock. Usually only bones, teeth, or shells are preserved, but occasionally impressions of softer parts can also be seen.

Imagine an ocean bed or a swamp floor in which new layers of sediment build up year after year over a long time. The layers of sediment preserve information about the past, just as tree rings record the history of a tree. Fossils preserved in the lowest layers in the sediment were made by plants and animals that lived a long time ago. Fossils in higher layers of sediment were made by plants and animals that lived more recently. If the plants and animals evolved during that time, we would expect to see changes in the fossils from the lower layers of sediment to the higher layers. This is exactly what scientists see.

As discussed in chapter 5, geologists have been systematically studying the earth since the 1600s, and they have become very skilled at figuring out both how and when different rocks are made. As they examine fossils from all over the world, certain patterns emerge. For example, in rocks over one billion years old they find only fossils of tiny, single-celled creatures. Fossils of fish are found in rocks 520 million years old and younger, fossils of amphibians in rocks about 380 million years old and younger, and so on. The fossil record shows the pattern of change over time.

> To see the ages of the earliest fossils of various plants and animals, refer back to the chart "Order of Creation in . . . Modern Science" in chapter 5 (p. 111).

The transitions among these types of fossils show evidence both of common ancestry and of the theory of evolution. The earliest amphibian fossils do not look like modern amphibians; they look like fish fossils of that same time period. Fossils of amphibians change over time to become more and more like modern amphibians. Similarly, the earliest reptile fossils look like amphibian fossils; reptile fossils change over time to become more and more like modern reptiles. The earliest bird and mammal fossils look like reptile fossils; bird and mammal fossils change over time to become more and more like modern birds and mammals. The theory of evolution makes predictions for how the fossil record will change over time and what transition fossils will be discovered. When scientists find new fossils, they use them to test the theory of evolution.

Let's look at the transition from reptiles to mammals. Mammals have three tiny bones in their middle ears called the hammer, anvil, and stirrup. These bones transmit vibrations from the ear drum to the inner ear. Reptile ears work differently; instead of these bones they have three extra bones in their lower jaw. As scientists look at transitional fossils leading from reptiles to mammals, they see an interesting transition in these bones. In the earliest fossils in this sequence, which are still mostly like reptiles and just a little like mammals, these three bones are still part of the jaw. In slightly later fossils these three bones are farther back in the head and neck of the animal. The bones still play a function in chewing, but they are also located in the region of the body that reptiles use for hearing. In still later fossils the other jaw bones of the animals are further modified so that the three bones of interest no longer play a role in chewing but still play a role in hearing.

In later and later fossils in the sequence, the three bones become increasingly like the middle ear bones of mammals.

> Whale fossils discovered over time can be used to test the theory of evolution. To read about an impressive series of transitional fossils, check out "Whales— Land or Sea Creatures?" on our website (www.faithaliveresources.org/origins).

The fossils that scientists have found so far are just a tiny fraction of all the plants and animals that have lived in the history of the earth. We don't have enough fossils to understand every detail of every change that happened. But we do have enough to give us the overall picture of what happened. Based on the theory of evolution, scientists have made predictions about what transitional fossils would look like and where they would probably be found. Over the last century many transitional fossils have indeed been discovered, confirming scientists' predictions. These fossils allow scientists to reconstruct a sort of family tree, indicating which species descended from which other species and when.

Comparative Anatomy Evidence

In addition to looking at the fossil record, scientists can look to other independent lines of evidence regarding evolution. One of these is comparative anatomy, in which scientists study the bodies of modern animals and look for similarities and differences across species.

Bats can fly, but they are mammals, not birds. Because they are mammals, the theory of evolution suggests that bats descended from non-flying mammals like rodents rather than from birds. This is confirmed by studying the bone structure in the bats' wings. The bone structure of bird wings is similar across all species of birds, including birds like the ostrich that can't fly. But the bone structure in bat wings is different; it looks much more like the front legs and feet of rodents like rats and mice. This makes

A Tree of Common Ancestry

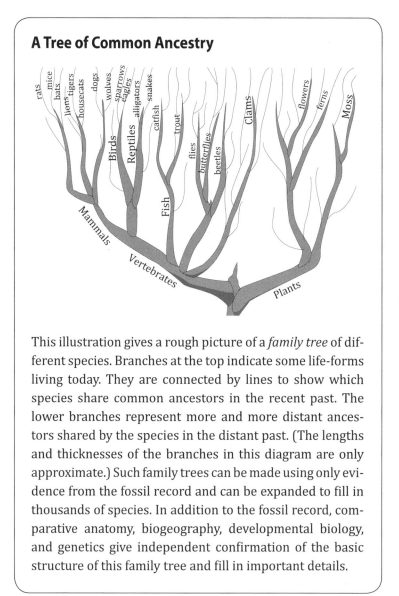

This illustration gives a rough picture of a *family tree* of different species. Branches at the top indicate some life-forms living today. They are connected by lines to show which species share common ancestors in the recent past. The lower branches represent more and more distant ancestors shared by the species in the distant past. (The lengths and thicknesses of the branches in this diagram are only approximate.) Such family trees can be made using only evidence from the fossil record and can be expanded to fill in thousands of species. In addition to the fossil record, comparative anatomy, biogeography, developmental biology, and genetics give independent confirmation of the basic structure of this family tree and fill in important details.

sense if rats, mice, and bats shared a common ancestor in the fairly recent past. The common ancestor of both mammals and birds lived much longer ago.

The retina is a collection of nerve cells in the back of the eye. Some cells detect light and send signals to other cells that process the signal before sending it to the brain. In all vertebrate animals that have eyes (fish, reptiles, amphibians, birds, mammals), the retinas are *inverted*; the cells that sense the light are located behind the cells that signal the brain. In invertebrate animals that have eyes (squid and octopus), the cells that sense the light are located in front of the cells that signal the brain. This organization makes sense if all vertebrate animals descended from a common ancestor that had inverted retinas, while invertebrate animals descended from a common ancestor that had non-inverted retinas. It might be argued that the non-inverted retinas of the squid and octopus are better for seeing things under water, but that wouldn't explain why vertebrates like fish that live underwater have inverted retinas.

These are just two examples. When scientists compare many different anatomical features across many different species, they reconstruct a family tree that looks very similar to the family tree from fossil evidence on page 197. The evidence from comparative anatomy strongly supports common ancestry and is consistent with the theory of evolution.

Biogeography Evidence

Biogeography is the study of how different species are distributed around the world. Large islands, especially those far from a continent, can make good biogeography case studies because they often have species that are not found anywhere else on earth. The Galapagos Islands are located about 600 miles off the west coast of South America. More than a dozen species of finches live on these islands. They resemble the finches living on the mainland, but the finches on the islands have a wide variety of shapes and behaviors. Some weigh as little as 10 grams when fully grown but others as much as 35 grams. Some species live on the ground, while others live in trees. Some species eat cactus, others eat seeds, and still others eat insects. One species of finch uses twigs or thorns to pry insects out of trees.

Scientists sometimes use the term *ecological niche* to refer to the manner in which a species fits into the ecosystem around it. On the Galapagos Islands many different ecological niches are filled by finches. On the South American mainland these same ecological niches are filled by a wider variety of birds. This raises questions:

▶ Why are these ecological niches filled by many bird types on the mainland but only by finches on the islands?

▶ Why are the finch species on the islands closely related to finches on the mainland, but even more closely related to each other?

If God specially created species to fill each ecological niche, no obvious reason for this particular pattern can be found. But if God used common ancestry and evolutionary processes to fill those ecological niches, the pattern makes sense. Some time ago a small group of finches flew from the mainland to the Galapagos Islands and stayed there. Their offspring evolved into more than a dozen different related species to fill different ecological niches.

One could argue that the finches on the Galapagos Islands are merely an example of microevolution (one species evolving a little bit and splitting into a few closely related species in order to adapt to a new environment). But this same pattern is seen again and again in much broader examples as well. Nearly all of the world's marsupials (mammals such as kangaroo and koalas that carry their young in pouches) live on Australia and its nearby islands. Marsupials fill many ecological niches; they're grassland grazers such as the kangaroo, tree-dwelling leaf-eaters such as the koala, burrowing carnivores such as the marsupial moles, and hunting carnivores such as the quoll. These species are significantly different from each other, but all are marsupials. On the other continents these ecological niches are filled by non-marsupial (placental) mammals. Why would all of these different ecological niches be filled by marsupials in Australia and the nearby islands but by placental mammals everywhere else? Common ancestry can explain it. The common ancestor of marsupials colonized

Australia long ago. Its descendants evolved into many species filling different ecological niches. On the other continents these ecological niches were filled by placental mammals.

For another example of how biogeography evidence supports common ancestry and the theory of evolution, check our website (www.faithaliveresources.org/origins) for "Biogeography of Fossils."

MODERN GENETIC EVIDENCE FOR THE THEORY OF EVOLUTION

The most dramatic new evidence for common ancestry and the theory of evolution comes from modern genetics. Scientists are now able to do detailed comparisons of genes between different species, something Darwin could not have anticipated.

In this section we'll discuss genetic evidence in some detail. Another modern line of evidence comes from developmental biology. Check our website (www.faithaliveresources.org/origins) for "Developmental Biology in Whales as Evidence for Evolution."

Genetic Similarity

When plants and animals pass on genes from parent to offspring, those genes are exact copies most of the time. Occasionally a mutation in a gene occurs. An organism with a mutated gene can pass on that mutation to its offspring. Scientists can use this fact to reconstruct family trees. Imagine three dogs—Fido, Rex, and Ace—all born from the same father and mother. In one particular gene Fido has a mutation that Rex and Ace do not. Fido passes on that mutation to his offspring, Flash and Zippy. In addition, suppose that Flash has a second mutation in that same gene and passes along both mutations to his offspring. In later generations a dog that has both mutations would be a descendant of Fido through Flash. A dog that has only the first mutation would be a descendant of Fido through Zippy. A dog with no mutation in that gene would be a descendant of Rex or Ace. No dog would have only the second mutation without the first mutation. Scientists are beginning to use these techniques to trace family trees of different purebred dog breeds. Dog breeds that are genetically similar have a more recent common ancestor than dog breeds that are genetically less similar.

Plants and animals have two versions of most genes, one from the mother and one from the father. If one of these versions has a mutation and the other does not, then an organism has only a 50 percent chance of passing on the mutated version to its offspring. This complicates the story somewhat, but scientists can deal with this by looking at many genes to reconstruct family trees. By looking at many genes, scientists can also deal with those rare instances in which a mutation in a later generation reverses a mutation from an earlier generation.

The theory of evolution predicts that this pattern extends to all species because all species share common ancestors. This is exactly the pattern that scientists see. The genes in lions, tigers, cougars, bobcats, and housecats are more similar to each other than to the genes in rats, horses, or other mammals. The genes in different species of rats, bats, and mice are more similar to each other than they are to genes in other mammals. This pattern also holds in other kinds of animals and plants. Genes in different species of sparrows are more similar to each other than they are to genes in any other types of bird. Genes in different species of trout are more similar to each other than they are to genes in any species of other types of fish.

Nested Pattern of Similarity

This pattern of *genetic similarity* also holds at higher levels in what scientists sometime call a *nested pattern* of similarity. Genes in all species of mammals are more similar to each other than they are to genes in any species of birds. Genes in all species of mammals and birds and reptiles are more similar to each other than they are to genes in any fish. Scientists can construct a tree of common ancestry of plants and animals based purely on the *genetic* similarities between species, without reference to fossils. Life-forms with the most similar genes had the most recent common ancestor, whereas life-forms with less similar genes had common ancestors longer ago. When this tree of common ancestry is compared with the trees found from fossils and from comparative anatomy, the trees agree with each other beautifully.

Genetic Diversity Within Species

One more piece of evidence for common ancestry comes from the *genetic diversity* in species. An *allele* is one of several different codes a gene might have. An organism has two copies of most genes, one from its father and one from its mother. If the two copies are identical (for example, two identical genes for black-colored fur), the organism is said to have a single allele for that gene. If the two copies are different (for example, one gene for black-colored fur and one for brown-colored fur), it's said to have two alleles for that gene.

If an entire species had descended from a single pair of animals in the recent past, the genetic diversity within the species should be limited. Genetic tests of many individuals in that species would find that most genes had at most four alleles, two from the original father and two from the original mother. (A few genes might have five or six alleles because of a mutation.) But scientists have discovered that plants and animals have much greater genetic diversity. Some genes have dozens of alleles, indicating that the "founders" of this species were more than just a single pair. Even more important, closely related species often share many of the same alleles, while more distantly related species share fewer of the same alleles. Common ancestry and the theory of evolution predict this pattern.

A more detailed explanation of genetic diversity and the theory of evolution can be found on our website (www.faithaliveresources.org/origins). Click on "Genetic Diversity Within Species."

COMMON FUNCTION OR COMMON ANCESTRY?

Opponents of the theory of evolution sometimes argue that different species share genetic similarities not because the species share a common ancestry but because the genes have a common function. Perhaps eagles, ravens, robins, and hummingbirds do not share a common ancestor, but instead their genes are similar to each other because all have similar body structures that allow them to fly. Perhaps this similarity of *function* requires that their genes also be similar. Each species may have arisen or been miraculously created individually, without a common ancestor, but they have genetic similarities to each other roughly in proportion to the similarities in their body structures and the functions that their genes perform. For genetic similarity, let's call this the *common function* theory, as opposed to the common ancestry theory. As we have seen, the common ancestry theory can explain the genetic similarities between species. The common function theory certainly can explain *some* of this genetic similarity, although it would have trouble explaining why the genes for bats are more similar to the genes of rats and mice than to the genes of birds.

But common function theory runs into trouble when trying to explain *pseudogenes*. Pseudogenes are broken or nonfunctional genes. A pseudogene looks like an ordinary gene but has one or more defects that prevent the body from using it to make molecules. It is like a page of a blueprint with a big "X" through it, indicating that the carpenters should ignore it. An ordinary gene can turn into a pseudogene if a mutation occurs that stops it from functioning. Ordinarily we would expect a mutation that turns a gene into a pseudogene to harm the organism. But when a species does not actually need a particular gene in order to survive, that gene can mutate into a pseudogene that can remain in the DNA of the species without harming it.

One example of a pseudogene comes from the gene necessary for the production of Vitamin C. Most mammals have a gene that allows them to make their own Vitamin C. In most mammals this gene is essential because Vitamin C is necessary for life. A mutation that made the Vitamin C gene nonfunctional would be fatal. Chimps, however, can survive without a Vitamin C gene because their diet includes a lot of fruit. Yet chimps do have a pseudogene for Vitamin C located in the same spot on the genome where most mammals have a functional gene for Vitamin C. The pseudogene has no function, yet it's in their genome. This makes sense if chimps share a common ancestry with other mammals. They inherited the Vitamin C gene from their distant ancestors, but sometime in the more recent past, perhaps about 8 million years ago, their ancestors had a mutation that turned it into a pseudogene. Because their ancestors were already living on fruit at the time, the loss of the gene's function was not fatal. If the common function theory were true, there would be no particular reason for pseudogenes to exist. The existence of pseudogenes, such as the one for Vitamin C, and their pattern in the DNA make sense if common ancestry is true.

Still more kinds of genetic evidence that favor common ancestry over common function can be found on our website (www.faithaliveresources.org/origins). Click on "Genetic Evidence for Evolution."

Maybe you've heard about the theory of evolution before and wondered about some of the following questions:

1. If the theory of evolution is correct, shouldn't there be half-cats/half-dogs alive today?
2. The second law of thermodynamics says that entropy (disorder) is always increasing. Doesn't this contradict the theory of evolution, which says that the orderliness and complexity in living organisms are increasing over time?
3. Doesn't evolution predict that changes in life-forms should be gradual? Aren't there big gaps in the fossil record where new life-forms suddenly appear?
4. Can evolution really produce big changes, such as changing fish into reptiles or reptiles into birds?
5. The first life-forms were simple, single-celled life-forms, but modern life is more complex. Can evolution explain how life got more complex over time?
6. Can evolution explain how life got started in the first place?

For a discussion of these questions visit our website (www.faithaliveresources.org/origins), "Questions Sometimes Asked About the Theory of Evolution." (The last two questions deal with the debate over Intelligent Design, the subject of ch. 10.)

THREE OLD EARTH VIEWS ON EVOLUTION

Now that we've summarized the scientific evidence for common ancestry and the theory of evolution, let's reconsider the old-earth creationist views, progressive creationism and evolutionary creationism. We will look at two versions of progressive creationism based on their different understanding of common ancestry.

▶ **Progressive Creationism without Common Ancestry**
Some progressive creationists believe that God did not use common ancestry. They believe that God acted miraculously many times in biological history to create new life-forms, so that later life-forms are not descended from earlier ones.

▶ **Progressive Creationism with Common Ancestry**
Others believe that God did use common ancestry but do not believe that the theory of evolution can explain all of the changes that took place in life-forms over time. They believe that while God used evolutionary processes to some degree, he also must have acted miraculously, perhaps on occasion modifying or adding new genes to certain organisms so that their offspring would begin to develop new capabilities.

▶ **Evolutionary Creationism**
Evolutionary creationists believe that God used common ancestry and also that he used the mechanisms of evolution to bring about changes in life-forms over time. They believe that while God could have chosen to act miraculously, he instead elected to design the mechanisms of evolution so that they would accomplish his goals in biological history through ordinary, natural means, under his ordinary governance, without the need for additional miracles.

It's likely that you'll encounter more than just two or three views on origins. For a summary of more than a dozen additional views, turn to "A Spectrum of Views on Origins" in the Appendix.

A vast amount of data that supports common ancestry has been collected over the past decades; we describe only a small portion of it in this chapter. Fossils, comparative anatomy, biogeography, and especially genetics provide independent, mutually supporting lines of evidence for common ancestry. Thus the scientific evidence is consistent with the second and third views described above, but it is difficult to reconcile with the first.

Of course, God *could* have specially created each type of plant and animal without using common ancestry. God could have created them with the same nested patterns of similarity across species that are predicted by common ancestry. That would have meant creating the anatomy of different species in the nested pattern, the fossil dates and geographical locations in a nested pattern, the genes in a nested pattern, and even pseudogenes in a nested pattern of similarity across all species. In other words, God could have specially created each type of plant and animal to *look as though* he had used common ancestry. But this idea runs into the same theological problems as the Appearance of Age Interpretation discussed in chapter 5. Since we know that God is the author of all truth, it would seem out of character for him to create each type of plant and animal through a miracle and then to put evidence into their DNA to suggest that he had created them through common ancestry.

The scientific evidence, while consistent with the theory of evolution, does not conclusively favor either the second or the third view. The evidence is consistent with both the Evolutionary Creationism view and the Progressive Creationism with Common Ancestry view. Differences between these views are discussed in the next chapter on Intelligent Design.

Comparing progressive creationism and evolutionary creationism leads us back to the issue of God's sovereignty over the natural world. The Bible teaches that God governs and is not absent from normal events in the natural world, and it teaches us that God can act miraculously when he chooses. This simple teaching removes much of the theological anxiety in the debate between these views. Remember that God's sovereignty is a foundation that allows us to examine the evidence in his book of nature to determine *how* God created.

QUESTIONS FOR REFLECTION AND DISCUSSION

1. This chapter lists several kinds of evidence for common ancestry and the theory of evolution. Which ones had you heard before? Which ones were new to you?
2. The Appendix describes a large spectrum of views on origins. With what range of views are you most comfortable?
3. In chapters 5 and 6 we discussed scientific evidence for great age of the earth, principles of biblical interpretation, and several different interpretations of Genesis 1. How do these relate to common ancestry and the theory of evolution?

ADDITIONAL RESOURCES

More on the scientific evidence for the theory of evolution presented from a Christian perspective:

Alexander, Dennis. *Creation or Evolution: Do We Have to Choose?* Kregel Publications, 2009.

Colling, Richard G. *Random Designer: Created from Chaos to Connect with Creator.* Browning Press, 2004.

Collins, Francis. *The Language of God: A Scientist Presents Evidence for Belief.* New York: Free Press, 2006.

Falk, Darrel. *Coming to Peace with Science.* Downers Grove, Ill.: InterVarsity Press, 2004.

Gray, Terry. "Biochemistry and Evolution," *Perspectives on an Evolving Creation.* Keith B. Miller, ed. Grand Rapids, Mich.: Wm. B. Eerdmans, 2003.

Harrell, Daniel M. *Nature's Witness: How Evolution Can Inspire Faith.* Nashville, Tenn.: Abingdon Press, 2008.

Lamoureux, Denis O. *I Love Jesus & I Accept Evolution.* Eugene, Ore.: Wipf and Stock Publishers, 2009.

Miller, Keith B. "Common Descent, Transitional Forms, and the Fossil Record," *Perspectives on an Evolving Creation.* Grand Rapids, Mich.: Wm. B. Eerdmans, 2003.

There are many essays on this topic available on the BioLogos website: biologos.org/.

More on the second law of thermodynamics and its relation to evolution:

Rusbult, Craig. "An Introduction to Entropy and Evolution: The Second Law of Thermodynamics in Science and in Young Earth Creationism" (www.asa3.org/ASA/education/origins/thermo.htm).

CHAPTER 10

INTELLIGENT DESIGN

Since very few politicians and judges are scientists, it might seem strange to ask for their opinions on scientific questions. But on October 6, 2005, a member of the press corps asked the White House press secretary, "Where does the president stand on the issue of Intelligent Design versus evolution?" And just a few days later, a fictional presidential candidate on the popular television show *The West Wing* was asked about his beliefs on these issues.

Why this media interest? Several months earlier the school board of Dover, Pennsylvania, had decided that a statement about Intelligent Design had to be read in ninth-grade science classes. A group of parents sued the school board, arguing that Intelligent Design theory is both unscientific and religiously motivated and therefore has no place in a public school science class. The judge ultimately ruled that the school board's required statement about Intelligent Design was an unconstitutional government endorsement of religion. Even before the judge issued the ruling, school board members who had supported the statement were voted out of office.

Within the last few years Intelligent Design (ID) has become a rallying point for battles over creation versus evolution. Intelligent Design does in fact make scientific claims that can be tested using the methods of science, but it also makes religious claims that cannot be tested scientifically. We will focus on two particular arguments:

▶ fine tuning
▶ biological complexity

But first let's take a look at the theory itself.

WHAT IS INTELLIGENT DESIGN THEORY ALL ABOUT?

Some of the conflict about Intelligent Design arises from confusing Intelligent Design *theory* with the Intelligent Design *movement*. Intelligent Design *theory* claims that there is evidence of design in nature and that the theory of evolution is inadequate to explain what is seen in the natural world. Although it is often motivated by religion, Intelligent Design theory avoids making specific religious claims. In contrast, the Intelligent Design *movement* is heavily influenced by religion and has political and cultural goals. It aims to combat the atheistic worldviews of naturalism (see ch. 2) and evolution*ism* (see ch. 8).

Opponents argue that Intelligent Design theory is not a scientific theory at all; it essentially means giving up trying to find a scientific explanation. They also argue that because supporters of Intelligent Design obviously believe that God is the Designer, the theory is a religious idea and therefore does not belong in the science classrooms of public schools. Supporters argue that Intelligent Design is a scientific theory based on scientific data. Their opposition to evolution might be religiously motivated, but they argue that any religious motivation is irrelevant. Intelligent Design, they say, can be studied and tested as a scientific alternative to evolution; therefore, it can be taught in the science classrooms of public schools.

What does *design* mean? All Christians—young-earth creationists, progressive creationists, and evolutionary creationists alike—believe that God designed the universe and all life. The universe and everything in it are not the result of some cosmic accident or some impersonal process. God created them intentionally. This shared belief does not make any particular claim about when or how God brought these things into existence. Saying that God designed the universe, the earth, and living organisms is another way of professing with all Christians throughout the ages the words of the Apostles' Creed: "I believe in God the Father almighty, maker of heaven and earth."

Within the last decade or so Intelligent Design *theory* has come to mean something much more specific. Intelligent Design theory focuses on two particular arguments. One argument is that the fundamental laws of physics and the basic parameters of the universe seem to be *fine-tuned* in order for life to exist. This indicates that a Creator designed this universe with the intention of bringing about life. (Fine-tuning is closely related to astronomy and the scientific ideas we introduced in ch. 7.) The second argument is that biological life is *irreducibly complex*, too complex to have evolved. An intelligent being must have intervened in some way during the history of life on earth in order to make life more complex. (Biological complexity is tied to evolution and the scientific ideas we introduced in chs. 8 and 9.) This biological complexity argument generates most of the scientific controversy and court battles on Intelligent Design. The fine-tuning argument receives much less press coverage.

Summary of the Fine-Tuning Argument

In chapter 7 we described some of the evidence that the universe is "fine-tuned" for life. When the fine-tuning argument is used to support a claim of Intelligent Design, it can be broken into three claims:

▶ *Scientific.* The fundamental laws of physics and the basic parameters of the universe fall within a narrow range of parameters that allow life to exist.

▶ *Philosophical.* No natural explanation exists for why the laws and parameters are tuned for life. Perhaps we just got lucky, but the most reasonable explanation is that the laws and parameters of this universe were designed for the purpose of supporting life. (The scientific data alone do not tell us who did the designing.)

▶ *Religious.* The best explanation is that God, the Creator revealed through Scripture, designed the laws and universal parameters from the beginning to bring about and support life.

The scientific claim is accepted by nearly all scientists of all worldviews, including atheists. The philosophical and religious claims are disputed. Most Christians agree that although fine-tuning doesn't *prove* that God exists, the *best* explanation for fine-tuning is that God designed the universe. Atheists and most agnostics prefer to explain fine-tuning by impersonal processes without reference to God.

BIOLOGICAL LIFE IS COMPLEX

Now we turn to the more controversial argument of Intelligent Design theory: biological life is too complex to have evolved according to the theory of evolution. Consider the amount of information contained in human DNA. If the DNA information in a single human cell were spelled out like letters on a page, it would take about one billion letters. The DNA letters spell out about 30,000 "words" called genes that are organized into 23 "books" called chromosomes. These letters are not arranged randomly. They encode information on how to make many kinds of molecules, all of which work together in complicated chemical reactions to keep your body healthy.

Consider the complexity of each individual cell. A cell is like a sports team in which individual players (the genes) work together to accomplish a goal (the functioning of the cell). But in this game there are 30,000 players! Each player has a unique position with a unique set of rules for how a player may interact with other players. Hundreds of coaches (other genes) continuously yell instructions to the players. In addition, the rules of the game keep changing subtly, depending on how many spectators are in the stadium (the environment around the cell). All of this organized complexity exists in a single living cell.

Consider the complexity in a single part of a single cell. Some single-celled bacteria have a *flagellum*. The flagellum, a whip-like structure the cell uses to swim, is a complex structure of more than twenty different types of protein molecules. If the gene for any one of those proteins is badly formed or missing, then the flagellum as a whole either functions much less efficiently or does not function at all.

Scientists believe that the very first life-forms on earth must have been much simpler than modern life-forms. Can simple organisms evolve into complex ones? Scientists have shown that the mechanisms of evolution can produce small changes over time (microevolution) and that modern organisms have descended from ancient ones (common ancestry). But it is not as clear exactly how simple life-forms can, over time, evolve into much more complex life-forms. Can animals without eyes evolve into animals that have eyes? Can bacteria that don't have flagella evolve into bacteria that do? Most scientists today are convinced that complex organisms can and have evolved from simpler ones, according to the theory of evolution. But advocates of Intelligent Design theory think that the mechanisms of evolution are not enough to explain all the complexity we see.

We can frame this question theologically. Did God design the laws of nature so that biological evolution is limited to making small changes? If so, it would be reasonable to suppose that God, at various points throughout biological history, must have miraculously superseded ordinary evolution in order to form more complex life. Alternatively, if God designed the laws of nature so that ordinary evolution *could* produce more complex life-forms, it would be reasonable to suppose that God used evolution under his providential governance without miracles to form more complex life.

Recognizing Design by Probability and Pattern

An object or an event is usually recognized to be *designed* when two things come together:

▶ *Probability.* It must be *very improbable* that the object or event could have existed without some intelligent being deliberately causing it.

▶ *Pattern.* The event or object follows a particular pattern that some intelligent being would reasonably want to create.

For example, imagine a Scrabble board with nineteen letter tiles lined up in a jagged row. If there were no recognizable *pattern*, you would probably assume that the letters had not been deliberately placed that way. If the first three letters happened to form a recognizable pattern, such as the word "can," you might still guess that the letters had not been deliberately placed, since it is not *improbable* for three out of nineteen letters to form a recognizable word. But if all nineteen letters formed a pattern, such as, "Can we play a game today" you would certainly conclude that the letters had been deliberately placed, since it's *very improbable* that nineteen letters would form a recognizable message unless someone designed it that way.

For another example of how probability and pattern come together in recognizing design, see "Probability, Pattern, and Design" on our website (www.faithaliveresources.org/origins).

Biological Complexity and Intelligent Design

Supporters of Intelligent Design theory use the ideas of *probability* and *pattern* to argue against the theory of evolution. Their claims can be phrased this way:

▶ First, it is highly *improbable* that complex biological features such as the bacteria flagellum could have evolved via the mechanisms of evolution.

▶ Second, it is reasonable to believe that some intelligent being would want to intervene during the history of life to create these complex biological *patterns*.

A lot of the debate has focused on the second claim and the nature of science. Some opponents say that Intelligent Design theory isn't scientific because it looks for a non-natural explanation. They say that science, by its very definition, must *always* look for natural explanations. Supporters of Intelligent Design say that this definition of science is too restrictive. They say that science tries to find the best explanations for the natural world, whatever those explanations might be. The divergence ends up being a debate about how people define *science*. This debate has political implications. If the second claim is considered *scientific*, then it can be discussed in the science classroom in public schools. If it is considered *religious* and not scientific, then Intelligent Design theory can be legally excluded from the science classroom in public schools.

It is difficult to test the second claim using the standard methods of science. But the first claim can be considered scientifically, and it must be shown to be valid before the second claim will be given a hearing. So we'll focus our attention on the first claim.

Is the Evolution of Complexity Improbable?

When the theory of evolution is explained in basic science textbooks or in popular books about science, it is often oversimplified with assumptions like the following:

▶ Each organism has a fixed number of genes.

▶ Each gene has a single function.

▶ A mutation in a gene only changes one DNA "letter" in the gene.

▶ The only way that a mutated gene can spread throughout an entire species is if that mutation offers some advantage (differential reproductive success) to the individuals who have it.

If these assumptions were true, then the first claim of Intelligent Design theory would almost certainly be true; it would be very improbable that complex things could evolve.

But these assumptions are false. Consider the following:

▶ The number of genes in an organism is not fixed. Entire genes or sets of genes can be duplicated. Some mutations can change large sections of a gene, not just a single letter.

▶ When a gene duplicates, one copy can keep its original function while the other copy can mutate to perform a new function. If the original gene already has two functions, each copy can mutate to specialize in one or the other function.

▶ Some functions can be accomplished by more than one set of genes. If a mutation significantly alters one gene, its functions might be accomplished by other genes so that the organism is not harmed by the mutation.

▶ Many mutations are *neutral*; they have no discernable effect yet can sometimes spread through a population. But after a gene accumulates several neutral mutations, one additional mutation can have a large effect.

▶ A mutation that is neutral in one environment can be beneficial in another environment.

This is a very incomplete list of some of the realities that scientists have discovered about evolution in the last several decades. Yet it is enough to show that evolution is more complicated—and more interesting—than the way it is sometimes presented in popular science books.

Given what scientists have discovered about evolution, is the evolution of complexity improbable? Scientists don't know; such a conclusion would be too difficult to reach based on the current level of technology and scientific knowledge. In a few cases scientists have already figured out a plausible explanation for how something complex could have evolved. One example is the evolution of the complex mammalian middle ear described in chapter 9 (p. 195). But in many other examples of complexity, scientists do not have enough information to decide whether its evolution was very probable or very improbable. It's too soon to say.

> A longer discussion of whether irreducible complexity can evolve, along with another example, can be found in two articles on our website (www.faithaliveresources.org/origins): "Is the Evolution of Complexity Improbable?" and "Ion Channels—An Example of How Complexity Could Evolve."

Summary of the Biological Complexity Argument

As with the fine-tuning argument, the biological complexity argument can be broken into three claims:

▶ *Scientific.* It is very improbable that complex biological features, such as the bacteria flagellum, could have evolved via the mechanisms of evolution. It is highly improbable that a living cell could self-assemble out of chemicals.

▶ *Philosophical.* If biological complexity could not have evolved and the complexity has an obvious function, then we should conclude that some sort of intelligent being intervened during the history of life on earth to create the complexity. (We cannot tell from the scientific data alone whether that intelligent being was God or "something" else.)

▶ *Religious.* The best explanation is that God, the Creator revealed through Scripture, intervened miraculously in the history of life to create biological complexity.

The scientific claim is disputed. Progressive creationists believe that it is true; evolutionary creationists that it is false. The majority of biologists believe that the general weight of evidence is against the scientific claim, in part because of all the genetic and fossil evidence for evolution in general (see ch. 9) and in part because there are a few specific cases in which scientists do understand how something complex could have evolved. But because there are many examples of biological complexity for which scientists do not yet have enough data to decide whether or not complexity could have evolved, the scientific question is still open.

Some critics of Intelligent Design say that the philosophical claim is just a religious claim in disguise. Supporters of Intelligent Design say that while the religious claim is fine for religious believers, it can be separated from the philosophical claim. Because the philosophical claim is religiously neutral and is based on scientific data, it can, they assert, properly be discussed in a science classroom.

BOTH EVOLUTION AND INTELLIGENT DESIGN COULD BE TRUE

The debate is almost always framed as though we have to make a choice between Intelligent Design or evolution; if one is true, then the other must be false. But perhaps both could be true.

Imagine that you are holding two plastic bags, each of which contains a collection of many tiny mechanical parts (springs, hooks, levers, gears, and so on). Suppose Bag A contains all the parts of a mechanical music box that has been disassembled. If you sealed this bag, you could shake it continuously twenty-four hours a day for years and years, but the music box would never reassemble. It would remain a bag full of separate pieces. Suppose that Bag B is similar to the first, except that all the pieces are designed to latch onto each other in specific ways. Springs, hooks, levers, and gears are designed so that when any two parts that belong together happen to collide, they latch together and stay latched together. In this case, the more you shake the bag the more the pieces come together in the right way. If you keep shaking the bag, eventually an entire functional music box will self-assemble out of all the component pieces. It's unlikely that anyone has made a self-assembling music box like this, but it probably could be done with enough time, determination, and cleverness.

Progressive creationists think of design like the music box parts in Bag A. First, all of the pieces of the music box are carefully designed so that they can fit together as a music box. (This is analogous to the idea that the fundamental laws and basic parameters of the universe seem to be fine-tuned so that atoms and molecules can exist and be assembled into living organisms.) Second, in order to become a functional music box, someone must put all the pieces together by hand. (This is analogous to the idea that biological complexity could not have evolved and must have been assembled through special miracles.) Once the assembly is done, the finished product is clearly a designed object.

Evolutionary creationists think of design as being like the music box parts in Bag B. The pieces of the music box must be very cleverly designed so that they fit together to make a music box *and* so that they can self-assemble in a shaken bag. Similarly, the fundamental laws of nature are fine-tuned so that life can exist and *also* so that life and biological complexity can self-assemble out of simpler pieces.

As noted earlier in this chapter, God could have made each star and each planet by a supernatural miracle. Instead, the natural world tells us that God fine-tuned the laws of nature so that stars and planets assemble out of simpler pieces through a beautiful system of regular natural processes that we can understand scientifically. Both progressive creationists and evolutionary creationists see this as evidence in favor of the idea that God designed the universe. Evolutionary creationists argue that God also fine-tuned the laws of nature so that simple organisms can evolve into complex ones and that this also should be seen as evidence *for*— not *against*—the idea that God designed living organisms.

What about the very first living thing on earth? Could it have evolved? Did God have to perform a miracle to create it? We explore these questions in "The Very First Living Cell" on our website (www.faithaliveresources.org/origins).

Intelligent Design Theory and "God of the Gaps"

Critics of Intelligent Design theory sometimes accuse it of being another example of "god of the gaps" thinking (see p. 49), which claims that the gaps in our scientific understanding are the best, or only, places to see God's hand at work. When science advances to fill the gaps in our understanding, the territory of this "god" shrinks. At its core Intelligent Design theory does have similarities with the "god of the gaps" idea. It argues that science cannot and never will be able to explain how the first living organism could self-organize or how biological complexity could evolve. This gap in scientists' ability to explain this is seen as evidence in favor of divine, miraculous intervention.

Supporters of Intelligent Design fall into the "god of the gaps" trap if they argue as follows:

If the theory of evolution is true and biological complexity can evolve, then atheists will have won that territory. We won't be able to point to living organisms as evidence of God's handiwork. Scientific evidence of God's existence and miraculous activity *must* occur somewhere in nature, and the complexity of life seems to be the best possibility. Therefore, biological complexity must be scientifically unexplainable.

They can avoid this trap if they instead argue as follows:

God could have chosen to create biological complexity any way he wished, whether by scientifically explainable processes or by miracles. Either way, God is in charge. But we think that it is *probably* true—for scientific reasons or for theological reasons or both—that God chose to use miracles. Therefore, we expect that scientists will never be able to find a satisfactory explanation for how biological complexity could have evolved.

Evolutionary creationists believe that God chose the ordinary operation of natural laws to govern the history of life on earth and to produce the world we live in today. This makes God's governance of the natural world appear hidden. Some evolutionary creationists find religious support for this idea in the theological concept of *kenosis*. For an explanation of this concept, see "Kenosis and Evolution" on our website (www.faithaliveresources.org/origins).

Public Schools, Public Debate, and the Intelligent Design Movement

This chapter has discussed particular arguments for and against Intelligent Design *theory*. But the subtleties discussed here are almost always missing in newspaper articles and TV reports about Intelligent Design. Even the quotes from leaders in the Intelligent Design *movement* and from their opponents do not explain these details. Sometimes reporters reduce the issue to a simple case of science versus religion, as though the two are fundamentally opposed. (See chs. 1 and 2 for our response to that simplification.) Often Intelligent Design theory is equated with young-earth creationism, ignoring the fact that most leaders of the Intelligent Design movement agree with the Big Bang and an old age of the earth.

In public debate the idea that something was designed by God is almost always set in *opposition* to evolution. For example, look back at the first paragraph of this chapter and the question "Where does the president stand on the issue of Intelligent Design *versus* evolution?" The question assumes that *either* the theory of evolution is true *or* the theory of Intelligent Design is true—we must choose one or the other.

This is a false premise. Evolutionary creationists argue that we don't have to choose between evolution and design. They argue that God designed the natural process of evolution to bring about plants and animals according to his plan. Evolutionary creationists are just as opposed to evolution*ism* and atheism as are the Intelligent Design and young-earth creationist movements, but they use different tactics. Instead of arguing against the theory of evolution, evolutionary creationists argue that scientific explanations like evolution reveal God's design rather than eliminate God's role.

Some of the leading advocates of Intelligent Design theory agree that evolution versus design is a false choice. They make the following distinction: If it turns out that biological complexity *cannot* evolve, then we have detected strong evidence that some sort of intelligent designer intervened during the history of life on earth. But finding that biological complexity *can* evolve would neither prove nor disprove that God designed living organisms but merely acknowledge that we cannot unambiguously detect it.

Unfortunately, these distinctions are usually lost in the media, in public debate, and in court cases. In these public forums evolution is often associated with atheism and God is associated with opposition to evolution. Science and religion are set up as enemies. As long as the public debate remains polarized and oversimplified, we will continue to have bitter debates about what should be taught in public schools.

Design for All Christians

Christians should not be satisfied with one side or the other of this false set of alternatives, presented as they are as mutually exclusive. As Scripture teaches we believe that God designed and created this world and everything in it. And when we look at the natural world through the eyes of faith, we see evidence of God's wisdom, creativity, and glory. But our belief does not rise or fall based on whether or not any particular scientific theory is true or false.

God has created an astounding universe filled with wonders, from the tiniest of particles to huge superclusters of galaxies, from the elegance of DNA to the complexity of ecosystems. The fundamental laws of physics appear to be fine-tuned for the exis-

On August 15, 2006, the *New York Times* reported that Harvard University had decided to give a team of researchers $1 million per year over the next few years to study whether or not it was possible that living cells could have self-organized on the early planet Earth, and if so, how they might have done so. One of the researchers is quoted as saying, "My expectation is that we will be able to reduce this to a very simple series of logical events that could have taken place with no divine intervention."

The very next day, Answers in Genesis, a young-earth creationist organization, denounced Harvard's plan on their website. The title of their article was "Harvard Allocates Millions to Prove There Is No God."

The quotations from both the Harvard researcher and Answers in Genesis present us with the same false dichotomy. Both imply that *either* there is a scientific explanation for how life first arose on earth *or* that God did it, as well as that any attempt to prove one is an attempt to disprove the other.

tence of stars and planets and molecules and living organisms. This does not scientifically prove that God exists, but it harmonizes with our belief that God created this universe as a fitting home for his many creatures, including human beings.

QUESTIONS FOR REFLECTION AND DISCUSSION

1. After reading this chapter, what do you think about the fine tuning of the laws of nature? What does it tell us about God? Should it be sufficient to convince someone that God exists?
2. If you think that God used miracles to create complex biological life, how do you avoid the "god of the gaps" problem?

3. If you think that God used natural evolutionary mechanisms to create complex biological life, how do you avoid the problem of deism?

4. How would you respond to someone who presented either the Intelligent Design theory or evolution as though you had to choose one or the other?

ADDITIONAL RESOURCES

More on the history of legal controversies over teaching creation and evolution in public schools in the United States:

Davis, Edward B. "Intelligent Design on Trial," in *Religion in the News,* Winter 2006, Vol. 8, No 3. (Available at www.trincoll. edu/depts/csrpl/RINVol8no3/intelligent%20design%20 on%20trial.htm).

Numbers, Ronald. *The Creationists.* Berkeley, Calif.: University of California Press, 1992.

More on arguments *for* Intelligent Design from a Christian perspective:

Behe, Michael. *Darwin's Black Box.* 2nd edition. Free Press, 2006.

Dembski, William, and Sean McDowell. *Understanding Intelligent Design.* Harvest House, 2008.

Meyer, Stephen. *Signature in the Cell: DNA and the Evidence for Intelligent Design.* HarperOne, 2009.

Strobel, Lee. *The Case for a Creator: A Journalist Investigates Scientific Evidence That Points Toward God.* Zondervan, 2005.

More on arguments *against* Intelligent Design from a Christian perspective:

Collins, Francis. *The Language of God: A Scientist Presents Evidence for Belief.* New York: Free Press, 2006.

Falk, Darrel. *Coming to Peace with Science.* Downer's Grove, Ill.: InterVarsity Press, 2004.

Gray, Terry, and Loren Haarsma. "Complexity, Self-organization and Design," *Perspectives on an Evolving Creation.* Keith B. Miller, ed. Grand Rapids, Mich.: Wm. B. Eerdmans, 2003.

Haarsma, Loren. "Is Intelligent Design 'Scientific'?" Lecture given at the American Scientific Affiliation annual meeting August 2005, available at www.asa3.org/ASA/PSCF/2007/PSCF3-07Haarsma.pdf.

More on scientific issues surrounding first life from a mostly neutral perspective:

Plaxco, Kevin, and Michael Gross. *Astrobiology: An Introduction.* Boston: Johns Hopkins University Press, 2006.

CHAPTER 11

SCIENTIFIC AND THEOLOGICAL ISSUES ON HUMAN ORIGINS

Questions about human origins are closer to the heart of Christian theology than questions about the age of the earth or even about the evolution of animals. In fact, concerns about human origins drive much of the debate over creation, evolution, and design.

Most theologians over the centuries have assumed that humans were specially created without common ancestry with animals. Yet the past century has seen tremendous growth in scientific information questioning this assumption. How strong is the scientific evidence? And, if humans *do* share common ancestry with animals, exactly which theological issues are at stake? In the past several decades an increasing number of Christian theologians and scientists have confronted these controversial issues. The church as a whole needs to be more aware of these developments. Because of the importance of these topics, the publishers of this book asked us to summarize the scientific and theological issues related to human origins.

We will first review the scientific evidence, taking a look at what God's revelation in nature seems to be saying. We will focus on three scientific areas:

▶ fossil evidence
▶ genetic similarity to animals
▶ genetic diversity in the human population

Then we'll look at four key theological issues related to human origins:

▶ the image of God
▶ the human soul
▶ original sin
▶ human mortality before the Fall

> If you have skipped ahead to this chapter, please go back and read the earlier chapters, especially chapters 2, 4, 5, 6, 8, and 9. These chapters lay the theological and scientific groundwork for this chapter and will help to avoid false impressions about the topic of human origins.

Another key issue is the interpretation of the Adam and Eve account in Genesis 2-3. We've set aside the entire next chapter for this topic. We will analyze five different scenarios regarding Adam and Eve, comparing in some detail how the theological and scientific issues discussed in this chapter play out in each scenario. As a preview of the next chapter, we summarize the five scenarios here. Please keep these in mind as you consider the issues discussed in this chapter.

Adam and Eve: Five Scenarios

▶ *Recent ancestors.* Adam and Eve were specially created about 10,000 years ago and were the first humans. All humans today have descended from them.
▶ *Recent representatives.* God created humans about 150,000 years ago, using progressive or evolutionary creation, and specially selected a pair of humans about 10,000 years ago to act as humanity's *representatives*. They chose to sin and their sinful status was applied to all humans.
▶ *Pair of ancient ancestors.* God used natural mechanisms to create pre-human *hominids*; then about 150,000 years ago God miraculously modified a pair of them into the first humans,

Adam and Eve. All humans today have descended from this pair.

▶ *Group of ancient representatives.* God created humans about 150,000 years ago, using evolutionary creation, and specially selected a particular group and revealed himself to them. They chose to sin and their sinful status was applied to all humans.

▶ *Symbolic.* God created humans about 150,000 years ago using evolutionary creation. No particular single event occurred in which all humans fell into sin at the same time, but many events happened in which various individuals and groups rebelled against God.

Many of the issues raised in this chapter and the next are areas of disagreement among Christians. For some of the most contentious issues, we summarize a range of answers Christian scholars have offered. We think that it's better for you to consider several possible answers—including ones with which we disagree—than for us to try to convince you that one particular answer is best. In some cases we are not fully satisfied with *any* of the suggestions we've heard.

SCIENTIFIC ISSUES ON HUMAN ORIGINS

In chapter 9 we discussed the origin of plant and animal species in light of multiple kinds of evidence: fossils, comparative anatomy, genetics, and genetic diversity. Scientists look at similar sorts of evidence to study human origins.

As in previous chapters we start by rejecting evolutio*nism*, an atheistic interpretation of the scientific data. The discovery of a scientific model for human origins would *not* eliminate God's action. In all the views presented in this chapter and the next, God is the Creator of humanity. The differences among the views lie in the area of *how* God accomplished this.

While Christians agree that God created humanity, they disagree on the mix of supernatural and natural processes he may have used. Christians hold three types of positions on this:

▶ *Special creation.* God acted miraculously to create the first human beings, independent of existing life-forms and without common ancestry with animals.

▶ *Miraculous modification.* God used progressive creation, including common ancestry with animals, to develop pre-human life-forms, after which he miraculously modified some of them to create the first human beings.

▶ *Evolutionary creation.* God created humans using both common ancestry and the mechanisms of evolution under his providential control but without miraculous action.

We will refer to these positions occasionally in this chapter and the next. Note that all three positions include the belief that God acted supernaturally when he revealed himself to human beings and established a spiritual relationship with us.

Fossil Evidence

It has been known for over a century that, when compared to that of all animals, human anatomy is most similar to that of chimpanzees and other apes. This knowledge led Darwin and other scientists to predict that if humans did share a common ancestry with animals, the most recent common ancestor would have been chimps and other apes. The hypothesis is *not* that humans descended from modern apes but rather that modern humans and apes share a common ancestor further back in the "family tree," a species that no longer lives today. In the same way modern lions, pumas, and house cats share a common ancestor that is extinct. If this hypothesis is correct, there should be fossils going back in time of human ancestors that look less and less like modern humans and more and more like the fossils of the ancestors of apes.

Over the last century several hundred fossils have in fact been found that fit this pattern (see "Hominid and Human Fossils,"

p. 234). Collectively these are often called *hominid* fossils. Many of them are only fragments of skeletons; some consist of multiple fragments found together; and a few include many parts of the skull or the entire skeleton. In this chapter we'll use the word *hominid* to refer to all of these types of fossils up to, but not including, modern humans (*Homo sapiens*).

Some of the oldest hominid fossils, called *Ardipithecus*, have brain sizes similar to those of modern chimps. The hominids that followed them looked more and more human. The *Australopithecus* hominids had slightly larger brains than *Ardipithecus*, and their skeletons imply that they walked upright. Fossils of *Homo habilis* show that they had still larger brain sizes, but still only half the size of those of modern humans. (Stone tools are sometimes found with *Homo habilis* fossils.) *Homo erectus* are still more modern-looking and had brains ranging from *Homo habilis* size to nearly modern human size (the largest actually fall within the range of modern humans). "Archaic" *Homo sapiens* look similar to modern humans but have some features closer to those of *Homo erectus*; their brain sizes fall within the range but average on the low side of those of modern humans. *Homo neanderthalensis* ("Neanderthals") were generally shorter and more heavily built than modern humans, but their brain size was similar to that of modern humans. (Good evidence shows that they made stone tools and a variety of other tools, controlled fire, and buried their dead.) When the brain size of hominid fossils is graphed as a function of time, it does not show abrupt jumps but rather a gradual increase from *Ardipithecus* to modern humans.

Note that scientists don't believe that *all* of these fossils are the direct ancestors of modern humans. Rather, they believe that these fossils are part of a family tree of which modern humans are one branch. For example, scientists believe that *Homo sapiens* did not descend from *Homo neanderthalensis*; rather, both are separately descended from an earlier hominid.

Hominid and Human Fossils

Ardipithecus	5 to 4 million years ago
Australopithecus	4 to 2.5 million years ago
Paranthropus	2.5 to 1.5 million years ago
Homo habilis	2 to 1.5 million years ago
Homo erectus	1.7 million to 250 thousand years ago
"Archaic" Homo sapiens	500 to 120 thousand years ago
Homo neanderthalensis	200 to 30 thousand years ago
Modern Homo sapiens	120 thousand years ago to present

Fossils of modern-looking *Homo sapiens* have been found dating back to about 120,000 years ago. Starting about 40,000 years ago, more extensive archaeological evidence shows *Homo sapiens* making tools out of stone, wood, bone, ivory, and antler and creating paintings and small sculptures. The oldest *Homo sapiens* fossils are found in Africa; later ones have also been discovered in Europe, Asia, and Australia. They are found in the Americas starting 12,000 years ago. Widespread evidence of agriculture and cities appears in the Near East around 7,000 years ago and more recently in other areas of the globe.

While fossils can tell us a lot about the anatomy of hominids, they cannot tell us as much about their behavior. Considerable debate occurs among scientists about the language abilities of *Homo erectus*, archaic *Homo sapiens*, and Neanderthals. Similarly, scientists debate whether Neanderthal burial practices or the paintings and sculptures created by modern-looking *Homo sapiens* more than 15,000 years ago held any religious significance.

Genetic Similarity to Animals

Scientists have found that human genes are very similar to chimpanzee genes, a little less similar to those of other apes, less similar to those of dogs and bears and other mammals, and still less similar to those of reptiles and birds. Thus, the evidence from human genes matches the predictions of common ancestry. Interestingly, scientists recently obtained some genetic material from Neanderthal fossils and compared it to human and chimp DNA. They found that Neanderthal DNA has a few differences from human DNA but that human and Neanderthal DNA are about eight times more similar to each other than they are to modern chimp DNA. This is what scientists should expect if humans and Neanderthals shared a common ancestor around 500,000 years ago.

As we discussed in chapter 9, it could be argued that the similarity between human and chimp gene sequences is explained by *common function* rather than *common ancestry*. Common function theory would argue that human and chimp gene sequences are similar only because they have similar functions in both species, not because of common ancestry. While common function theory might explain some similarities in gene sequences, it has trouble explaining other evidence such as similarities in pseudogenes.

In chapter 9 we mentioned that most mammals have genes that allow their bodies to make Vitamin C. But in chimps and most apes the gene for Vitamin C is broken and nonfunctional— a pseudogene. It turns out that humans also have a Vitamin C pseudogene instead of a functioning Vitamin C gene. This is why humans, like chimps and other apes, need to regularly eat foods that contain Vitamin C. On the basis of the common function theory, it is hard to explain why humans, chimps, and other apes would have a non-functional pseudogene in the same location that other mammals have a functional Vitamin C gene. But on the basis of common ancestry theory, the existence and location of the Vitamin C pseudogene makes sense.

For more genetic evidence that favors common ancestry over common function, see "Human Genomic Organization and Introns" on our website (www.faithaliveresources.org/origins).

Francis Collins, head of the Human Genome Project, one of the leading authorities on human genetics, and an evangelical

Christian, describes the genetic evidence for common ancestry in more detail in his recent book *The Language of God* (2006). He concludes, "The study of genomes leads inexorably to the conclusion that we humans share a common ancestor with other living things."

Genetic Diversity in the Human Population

Scientists can study the genetic diversity within a species by looking at the different versions of a gene (*alleles*) that exist in the population today (see ch. 9, p. 202). In the human population the genes with the most diversity are in a section of DNA called the *histocompatibility complex* that includes genes important for the immune system. Some of these genes have more than 150 alleles in the human population. Assuming that natural mutation rates in the past are similar to mutation rates today, this number of alleles is far greater than would be expected if all humans were descended from a single pair that lived only 10,000 years ago. Even if the first pair had lived hundreds of thousands of years ago, natural mutation rates are not fast enough to produce this many different alleles from a single pair of ancestors.

Population geneticists build mathematical models to reconstruct the history of populations of animals, plants, or bacteria, taking into account reproduction and mutation rates. When they look at the genetic diversity in the human population today, their best mathematical models indicate that the ancestors of humans were at a minimum number (called a *population bottleneck*) about 150,000 years ago. The size of this bottleneck was about 10,000 to 100,000 individuals. The genes in all humans today would be descended from that common set of founders. If God created humanity using *evolutionary creation* without miracles, these models indicate that humans today descend from this larger group rather than from a pair of individuals.

You may have heard scientists talk about a *genetic Adam* or a *mitochondrial Eve*. Perhaps you've wondered whether these could be connected to the biblical Adam and Eve. For an explanation for why scientists *don't* believe that all humans descended from these two individuals, see "Genetic Adam and Mitochondrial Eve" at our website (www.faithaliveresources.org/origins).

When scientists build models, as best they can, combining the fossil and genetic evidence, they conclude that this population bottleneck occurred in Africa about 150,000 years ago. After that point the population increased and humans spread into Europe and Asia, then to Australia and the Pacific islands, and finally to the Americas around 12,000 years ago.

THEOLOGICAL ISSUES ON HUMAN ORIGINS

A number of important theological issues regarding human origins deserve our attention. In this section we'll focus on four issues: the image of God, the human soul, original sin, and human mortality before the Fall.

Humans Are God's Imagebearers

For centuries theologians have considered what exactly the Bible means by saying that humans are created "in the image of God." They don't all agree. Theologians have given three general answers, identifying the *image of God* with

▶ *our mental and social abilities.* Some have looked at the differences between animals and humans and have identified the image of God in terms of our greater mental and social abilities. Humans are superior to animals in intelligence, rational thinking, language use, creativity, the ability to build social relationships, and so on. Humans share these characteristics of God to a far greater extent than animals do.

▶ *the personal relationship between God and humans.* Others have identified the image of God with God's choice to have a personal relationship with us. God has revealed himself to us, holds us morally accountable for our actions, and intends for us to live with him eternally. It is because of this relationship that we carry the image of God.

▶ *our commission to be God's representatives and stewards.* Others have identified the image of God with our commission from God to be his representatives and stewards in this world. In ancient Near Eastern cultures a king would put statues (*images*) in distant parts of his realm to indicate his sovereignty. God transformed this idea by declaring humans to be his *living* images on earth to represent his sovereignty and to act as his stewards. This understanding of *image* fits with the second commandment, in which God forbids the making of graven images of himself; humans are already his living images.

How do these views of the image of God relate to common ancestry and evolution? Is it possible that God's imagebearers evolved from simpler life-forms?

If being made in God's image is about our abilities, then yes. God could have given us our mental and social abilities purely miraculously, through *special creation* of the first humans. Or God could have given us those abilities through a combination of natural and miraculous processes, through *miraculous modification* of pre-human hominids. Or he could have given us those abilities simply through his governance of natural processes, through *evolutionary creation.* Either way, our mental and social abilities are a gift from God; they are part of God's intention for us. Thus this aspect of the image of God is not at all denied by common ancestry with animals or by evolution. Our status as imagebearers is not based on *how* we got these abilities but on the fact that they are a gift from God and are integral to God's plan.

What about the second and third views on the image of God? These focus on our spiritual relationship and status before God, relying on his supernatural action. At some point in human history God chose to establish a relationship with human beings and declared them to be his imagebearers. God did this uniquely with human beings, not with any animal species with whom we might share common ancestry. This supernatural act of God is independent of how we received our physical and genetic characteristics or our mental and social abilities. Thus, these aspects of the image of God are also independent of whether or not we share common ancestry with animals.

The Human Soul

The evidence for human evolution also raises questions about the human soul. Although no single definition of the word *soul* satisfies all theologians, a general consensus is that our souls are the *immaterial* parts of ourselves. In some sense our souls are the essential core of who we are as individual persons, our mental and spiritual selves.

> Some Christian traditions talk about humans as three-part beings: body/mind/soul or body/soul/spirit. In this book the word *soul* refers to the combination of mental and spiritual aspects.

Through our souls we have the ability to know, love, and respond to God and to be his imagebearers. By God's grace our souls survive when our bodies die, and God promises his children new life in resurrected bodies in the new creation.

People have developed many theories to attempt to explain how the soul relates to the body. For example, the worldview of atheistic materialism claims that humans are nothing more than bodies with mental capacities; no soul survives after the death of the body. Other worldviews claim that our material bodies are unimportant and that our immaterial souls are naturally immortal; our souls are actually better off after death when they are no longer tied to bodies. Christianity rejects both of those theories; in the words of the Apostles' Creed, we believe in "the resurrection of the body and the life everlasting."

Even within Christian tradition, a range of theories exists for how the soul relates to the body. We'll briefly summarize three of the more common theories to see how they connect to different views on human origins.

▶ **Theory 1:** The body and the soul are two different entities, one material and the other immaterial, conjoined by God to make one person. The body without the soul is dead. The soul can exist without a body but in a diminished state.

▶ **Theory 2:** The body is material and the soul is immaterial, but they should not be thought of as two different entities. Matter is an ingredient, not a separate entity. The soul organizes and empowers the body, endowing it with its essential human characteristics, such as self-consciousness, reason, will, and the ability to relate to God. The soul can exist without a body but in a diminished state.

▶ **Theory 3:** Our bodies, in particular the functioning of our brains, give rise to all of our mental abilities, including our capacity to have personal relationships with other humans and with God. But our spiritual life also depends on God supernaturally establishing a relationship with us, revealing himself to us, and making promises to us. Because our mental and spiritual capacities are so dependent on our bodies, disembodied souls cannot exist without God's miraculous, sustaining activity.

All three theories affirm the biblical view of the importance of the body, both for life now and ultimately for everlasting life in resurrected bodies. All three agree that our life now, our survival after death, and ultimately our life in the new creation are possible only by God's grace.

How do these theories relate to views on human origins? Theories 1 and 2 hold that God performed some sort of miracle to create the first human souls. Therefore they are usually associated either with the view that God specially created the first humans or with the view that God miraculously transformed pre-existing hominids to create the first humans. Although less common, it is also possible to combine these theories of the soul with the view that God created the first humans' physical and mental characteristics through the natural mechanisms of evolution.

The third theory about the soul does not require a miracle for the origin of human mental abilities; these could have developed through evolutionary processes. For that reason this theory is most often associated with evolutionary creation of humans. But we should note that this theory about the soul still entails God's supernatural activity in revealing himself to humans and establishing a relationship with them. Although less common, it is also possible to combine this theory of the soul with the views that God created humans through special creation or miraculous transformation.

Original Sin

The topic of original sin is closely related to that of Adam and Eve. It involves three main issues:

▶ The *situation* of original sin. Are babies born sinful, or are they born with a blank slate and fall into sin later?

▶ The *transmission* of original sin. How is the sinful nature passed through the generations?

▶ The *historical origin* of original sin. When was the first time that human beings sinned?

The situation of original sin

In the early Christian church some leaders taught that humans are born in a state of natural righteousness—that is, without sin. They said that humans are, in principle, capable of living sin-free lives and of achieving righteousness without the saving act of Christ; humans simply learn sin from the bad example of society. This view came to be called *Pelagianism* and was condemned as a heresy by Augustine and other church leaders around A.D. 400.

The doctrine of original sin was formulated, in part, in response to Pelagianism. The doctrine of original sin says that because of humanity's rebellion against God all human beings are sinful from birth and have no ability to be righteous apart from Christ's redemption (Rom. 3:22-24). All humans, including infants, need Christ's righteousness.

The doctrine of original sin upholds salvation as an act of God's grace that no human can earn. All five scenarios for Adam and Eve summarized on pages 230-231 and discussed in chapter 12 are compatible with the view that humans today are born sinful and cannot be righteous apart from Christ.

The transmission of original sin

Theologians differ on the question of how original sin is passed on through the generations from Adam and Eve to us today. It could be transmitted

- ▶ spiritually.
- ▶ socially.
- ▶ biologically.

Some suggest that original sin applies to all people because it is primarily a *spiritual status* before God. The disobedience of Adam and Eve put all humans, including infants, in a *state of sin* before God; they lost the *state of grace* that should exist between God and humans. The spiritual fellowship between God and humanity was broken, and it is impossible for us to restore that relationship by our own efforts. Because of this loss of fellowship with God, each of us inevitably commits sinful acts. This spiritual status is shared by all humanity. Many times in the Bible God condemns nations or humanity as a whole for their corporate guilt. Original sin is more than individual acts committed by an individual person; it is the corporate human condition.

Other theologians, while not denying the broken spiritual relationship, suggest that sin is transmitted from one human to another through *social interaction* and imitation. Humanity's sinfulness makes it impossible for us as individuals to avoid sin or its consequences. Each one of us in turn individually contributes to humanity's sinful condition. Whatever we do, for good or ill, inevitably affects others. Our sinful acts contribute to the sinfulness of others; children inevitably learn to sin by imitation.

Still other theologians emphasize the *biological* aspect of the transmission of original sin. Physically, and even genetically, humans are prone to sin. From our everyday experience we know that sin can become habit, and a bad habit increases the temptation to sin more. Our predisposition to sin is partly learned behavior. It is also partly genetic; each of us is born with biological predispositions to certain sins, whether that be pride, a nasty temper, or alcohol abuse. Some say that a predisposition to sin is biologically hardwired into human beings. To the extent that predispositions are genetic, children inevitably inherit a tendency to sin.

These views of the transmission of original sin are not necessarily contradictory. Many theologians say that all three aspects—spiritual, social, and biological—are part of the transmission of original sin. As we'll discuss in the next chapter, these three aspects of the transmission of original sin play out somewhat differently in our five scenarios about Adam and Eve.

The historical origin of original sin

When did the first sin occur? What was the spiritual state of humanity beforehand? In the first four scenarios with Adam and Eve as ancestors or as representatives, humanity disobeyed a clear command from God at one particular moment in history and fell into sin. In the scenario in which Adam and Eve are symbolic, humanity fell into sin through a series of disobedient acts. All agree that humans fell into sin, but they disagree about the status of humanity *before* the first historical sin.

In the scenarios with Adam and Eve as recent or ancient *ancestors*, Adam and Eve are viewed as living in a state of *original righteousness*—in full communion with God—before they disobeyed him. (See, for example, the Westminster Confession, Article VI.2; www.reformed.org/documents/wcf_with_proofs/.) God had revealed himself to them, and Adam and Eve were in righteous fellowship with God before they broke that relationship. The scenarios of Adam and Eve as recent or ancient *representatives* raise the question of the spiritual status of the *other* humans living in the time before Adam and Eve sinned. According to the symbolic view of Adam and Eve the state of original righteousness was a potential state, not an actual state. It is a state that humanity might have achieved had it responded obediently to God's revelations. (These scenarios are discussed in more detail in the next chapter.)

Human Mortality Before the Fall

Was there death before the Fall? All old-earth views imply millions of years of animals living and dying before humans lived. Some of the scenarios of Adam and Eve also have humans dying before the Fall, which conflicts with a common interpretation of Scripture that says that physical death is a consequence of humanity's fall into sin. We'll consider animal death first, then human death.

Physical death of animals

Scripture passages that discuss death as a consequence of sin (Gen. 2:16-17; 3:19, 22; Rom. 5:12-21; 1 Cor. 15) clearly refer to humans, but it is uncertain whether they also refer to animals. Because of this ambiguity theologians debated the question of animal death for centuries before modern science. Some said that animals would have been immortal if Adam and Eve had not sinned. They point to prophetic passages like Isaiah 11:6-7 and 65:25 that refer to predatory animals like bears and lions living peacefully with cows and lambs. These passages clearly refer to the new heaven and the new earth, but some theologians say that they also describe the earth before human sin. Others have argued that these verses should only be applied to the new heaven and the new earth and do *not* refer to the first earth or the state of animals before sin. They suggest that a limited lifespan and physical death are a natural part of animal existence. They point to verses like Job 38:39-40 and Psalm 104:21 that refer to God providing prey for predatory animals, implying that God intended from the beginning for animals to die and make way for new generations of creatures.

Since Scripture itself is unclear on the original state of animals, and theologians have offered competing interpretations that can be reconciled with Scripture, this seems to be a situation where we can look to God's revelation in the book of nature to help us decide on the best interpretation of Scripture. The abundant scientific evidence for an old earth and the long history of life on earth clearly indicates that death was a natural part of animal existence from the beginning.

Human death

On the subject of human death Scripture is less ambiguous but still open to multiple interpretations. Some have argued that the passages that identify death as a consequence of sin (such as Rom. 5:12-21 and Gen. 2:17) refer only to *spiritual* death (separation from God), not to physical death. They point to Genesis 2:17 ("when you eat of it you will surely die . . .") and note that Adam and Eve did *not* physically die immediately after disobeying God, although they were immediately separated from God. They also point to 1 Corinthians 15:56 ("The sting of death is sin") and argue that sin and separation from God are the real enemies, not death itself. Human physical death, they argue, was part of our original created physical nature.

Others have interpreted Scripture to mean that physical death as well as spiritual death is a consequence of sin. This view has been more common throughout the history of the church. In 1 Corinthians 15, a chapter about the physical resurrection of Christ and the *physical* resurrection of the body for those who die in Christ, Paul states that death came through the sin of Adam (vv. 21-22). Verse 26 speaks of physical death not as a good part of God's original plan but as "the last enemy to be destroyed."

Within this view (that the fall caused both physical and spiritual death) there is still disagreement about the original physical state of humans. Was physical immortality built into human bodies from the beginning, or was immortality a *potential* gift that humans could have received from God only if they had chosen not to sin? If created physically immortal, Adam and Eve would have had bodies that did not age and that overcame all disease and injury. If they only had the *potential* for immortality, Adam and Eve would have had bodies similar to our own that could become immortal only by God's miraculous action. Genesis 2-3 does not explicitly teach that Adam and Eve were created with physical immortality. In fact, the presence of the *tree of life* in the Garden of Eden suggests otherwise. The tree of life was in the garden before Adam and Eve sinned (Gen. 2:9). What was the purpose of the tree if Adam and Eve were already physically immortal? The presence of the tree makes the most sense if Adam and Eve were mortal and needed divine action to make them immortal. The tree seems to represent a *potential* of immortality, a gift of God that was lost to humanity when they sinned: "He must not be allowed to reach out his hand and take also from the tree of life and eat, and live forever" (Gen. 3:22). The tree of life appears again in Revelation 22 in the New Jerusalem on the new earth, where death has been abolished and all of God's people have been raised imperishable. Through Christ, God's plan of human immortality is finally fulfilled in the new creation.

> For various interpretations of the tree of life, see "Three Interpretations of the Tree of Life" on our website (www.faithaliveresources.org/origins).

On a purely physical level it is difficult to imagine how human bodies would function if they had built-in physical immortality. Our bodies today grow old and respond to disease and injury much like those of other creatures. The human immune system for fighting disease is very similar to that of animals; it has a limited ability to repair injury. A physical body that could recover from *any* injury or disease would have had to be constructed on entirely different principles from our current bodies. Aging itself is built into our DNA; our very cells wear out and die. It would have taken more than a few minor changes to make our human bodies immortal—it would have required a complete reconstruction with fundamentally different chemical and biological processes. In fact, there seems to be no physical object in this created universe that is built to last forever. Mountains erode, continents shift, and our own sun does not have enough fuel to shine forever. The natural processes God designed do not seem to support immortality in *this* creation. True immortality seems to be something for the next creation, made possible only by God's miraculous and gracious action.

RECONCILING THE ISSUES

In this chapter we've raised many scientific and theological issues relating to human origins. You might be wondering whether it's possible to tie these all together. In the next chapter we'll try to do just that as we discuss how each issue plays out in five different scenarios for Adam and Eve. Although the truth within this complicated topic is not always clear to our limited human understanding, we can take hope. Our hope is in God's sovereignty and in his character. God is the author of all truth in both nature and Scripture.

QUESTIONS FOR REFLECTION AND DISCUSSION

1. This chapter summarized some of the fossil and genetic evidence on human origins. What had you heard before? What was new to you?
2. How would you describe what it means for humans to be created in the image of God?
3. How would you define the soul? How do you think the body and the soul are related?

4. What view or views do you think best describe how *original sin* is passed from one generation to the next?
5. In your view, what does Scripture teach regarding the immortality of humans before the Fall? Was physical death part of God's plan for humanity?

ADDITIONAL RESOURCES

More on scientific evidence regarding human evolution, from a mostly neutral perspective:
Heslip, Steven. "Time-Space Chart of Hominid Fossils." www.msu.edu/~heslipst/contents/ANP440/.

More on scientific evidence regarding human evolution, from a Christian perspective:
Collins, Francis. *The Language of God: A Scientist Presents Evidence for Belief.* New York: Free Press, 2006.

Hurd, James. "Hominids in the Garden," *Perspectives on an Evolving Creation.* Keith B. Miller, ed. Grand Rapids, Mich.: Wm. B. Eerdmans, 2003.

Stearley, Ralph. "Assessing Evidences for the Evolution of a Human Cognitive Platform for 'Soulish Behaviors,'" *Perspectives on Science and Christian Faith,* Vol. 61, p.152-174, September 2009.

Venema, Dennis R. "Genesis and the Genome: Genomics Evidence for Human-Ape Common Ancestry and Ancestral Hominid Population Sizes," *Perspectives on Science and Christian Faith,* 62:166, September 2010.

Wilcox, David. "Finding Adam: The Genetics of Human Origins," *Perspectives on an Evolving Creation.* Keith B. Miller, ed. Grand Rapids, Mich.: Wm B. Eerdmans, 2003.

More on theological issues of human evolution:

Collins, Robin. "Evolution and Original Sin," *Perspectives on an Evolving Creation.* Keith B. Miller, ed. Grand Rapids, Mich.: Wm. B. Eerdmans, 2003.

Lamoureux, Denis O. *I Love Jesus & I Accept Evolution.* Eugene, Ore.: Wipf and Stock Publishers, 2009.

Murphy, George. "Roads to Paradise and Perdition: Christ, Evolution, and Original Sin," *Perspectives on Science and Christian Faith.* 58:109, June 2006.

Young, Davis. "The Antiquity and the Unity of the Human Race Revisited," *Christian Scholar's Review.* XXIV:4, May 1995. Available at www.asa3.org/asa/resources/CSRYoung.html.

CHAPTER 12

ADAM AND EVE

T he Apostles' Creed doesn't mention Adam and Eve, but it does talk about *the forgiveness of sins*. Theologians throughout church history have reflected on human sinfulness and mortality in light of Genesis 2 and 3. Many paintings have depicted the temptation of Adam and Eve and their expulsion from the garden. Usually they include a snake, the tree of the knowledge of good and evil, and its fruit, often shown as an apple. While theologians disagree on whether these elements should be considered literally or symbolically, all agree that they represent important spiritual realities.

As we learn more about human history from archeology and studies of genetics, more questions arise. How long ago did Adam and Eve live? Did all people descend from them, or did other humans live before them? Are Adam and Eve merely symbolic of early humanity? Christians have struggled with these questions for centuries and have considered many different answers. In this chapter we'll discuss five scenarios (briefly summarized in ch. 11, pp. 230-231) that illustrate the range of answers we've heard. For the sake of clarity we've limited the list to five, although intermediate positions are also possible. We'll discuss Adam and Eve as

▶ recent ancestors.
▶ recent representatives.
▶ a pair of ancient ancestors.
▶ a group of ancient representatives.
▶ symbolic.

Theologians in all five scenarios agree that God created humanity in his image. God revealed himself to them, began a relationship with them, and gave them moral and spiritual obligations. They chose to sin, and that sin was transmitted to the rest of humanity.

> If you have skipped ahead to this chapter, please go back and read the earlier chapters, especially chapters 2, 4, 5, 6, 8, 9, and 11. These chapters lay the theological and scientific groundwork for this chapter and will help to avoid false impressions about the topic of human origins.

Each of the five scenarios about Adam and Eve faces significant challenges from God's revelation in nature, God's revelation in Scripture, or both. In this chapter we'll look at each scenario in more detail in light of the scientific and theological issues raised in chapter 11. (See a summary of these issues below.) Keep in mind that our goal here is to promote informed discussion in the church rather than to defend one particular position.

Summary of Scientific and Theological Issues
Scientific Issues

▶ *Fossil evidence:* Scientists have found fossils of hominids going back five million years and of modern-looking humans going back at least 120,000 years.

▶ *Genetic similarity to animals:* Similarities between human and animal genetic sequences support common ancestry.

▶ *Genetic diversity in the human population:* The diversity in the gene pool is much more than would be expected if all were descended from a single pair.

Theological Issue: The Image of God

▶ View 1: *Social abilities:* We have mental and social abilities far above those of animals.

▶ View 2: *Personal relationship:* God chooses to have a personal relationship with us.

▶ View 3: *Representatives and stewards:* God commissioned humans to be his representatives and stewards in this world.

All three views are compatible with each other and with all five scenarios about Adam and Eve.

Theological Issue: The Human Soul

▶ *Theory 1:* The soul and the body are two different entities (one immaterial, one material). The soul must have been created miraculously.

▶ *Theory 2:* The body is material and the soul is immaterial, but they should not be thought of as two different entities. The soul organizes and empowers the body, endowing it with its essential human characteristics, such as self-consciousness, reason, will, and the ability to relate to God. The soul must have been created miraculously.

▶ *Theory 3:* The soul consists of our mental and relational abilities (arising from our bodies) plus God's spiritual relationship with us. God could have created the soul through the natural mechanisms of evolution plus his special revelation to us.

All three theories are compatible with all five scenarios about Adam and Eve.

Theological Issue: Original Sin

▶ *Situation of original sin:* No one can be righteous apart from Christ.

- ▶ *Transmission of original sin:* Sin is transmitted to other humans biologically, socially, as a spiritual status, or all three.
- ▶ *Historical origin of sin:* Was it a single act or multiple acts? Was original righteousness before the Fall an actual or a potential state?

Theological Issue: Human Mortality Before the Fall

- ▶ View 1: The Fall resulted in spiritual death only; humans were naturally mortal before the Fall.
- ▶ View 2: The Fall resulted in both spiritual and physical death; humans were naturally immortal before the Fall.
- ▶ View 3: The Fall resulted in both spiritual and physical death; humans were naturally mortal but potentially immortal before the Fall.

FIVE SCENARIOS ABOUT ADAM AND EVE

Adam and Eve as Recent Ancestors

In this scenario God *specially created* a pair of humans named Adam and Eve about ten thousand years ago. They were the first humans, and no other humans lived at that time or earlier. All humans today have descended from them, inheriting their sinful status.

Through most of church history most Christians have held this view. It allows a fairly literal interpretation of Genesis 2 onward. Because a great deal of what theologians have written about original sin throughout the centuries assumes this scenario, it's the easiest one to deal with theologically. It is compatible with all the traditional views of the soul and the views of humans as God's imagebearers. It is compatible with each of three views of human mortality before the fall, and it fits with traditional Protestant and Roman Catholic views on original sin: Adam and Eve started in an actual state of original righteousness and fell into a state of sin through an act of disobedience; original sin is transmitted to their descendants biologically, socially, and/or spiritually.

This scenario is not completely without theological problems. For example, who was Cain's wife? Who was Cain afraid would kill him? Also, if Genesis 4 is read in the same literal fashion, the text suggests that many cities of large population were built just a few generations after Adam. Where did these people come from? Usually these problems are dealt with by assuming that Adam and Eve had many children beyond those mentioned in Scripture and that the genealogies in Genesis only list representative members, skipping many generations.

This scenario is the most difficult one to reconcile with the evidence God gives us in nature. Archaeological evidence, such as tools, artwork, and indications of fire use, shows humans living on every continent, including North and South America, for more than 10,000 years. These peoples could not all have descended from a single pair living in the Near East about 10,000 years ago; there would not have been sufficient time for humanity to spread around the globe. In addition, fossils that look just like modern humans have been found in Asia and Europe dating back more than 30,000 years and in Africa over 100,000 years. If God made the first pair of humans only 10,000 years ago, why did he also create—either miraculously or via evolutionary mechanisms— all of those other *apparently* human creatures over the preceding 100,000 years?

Several lines of genetic evidence challenge this scenario. It has difficulty explaining why humans have many of the same non-functional pseudogenes as chimps and other apes. It also has difficulty reconciling the genetic diversity in the human population today, where some genes have more than 150 *alleles*. If God started by specially creating a pair of humans—who would together have at most four alleles of each gene—he would also have had to miraculously increase the number of mutations in these genes over several later generations in order to produce the genetic diversity we see in the human population today.

> The recent ancestors scenario has particular difficulty explaining why the *introns* in human DNA are so similar to the introns in chimps and other apes and why the human genome is organized so much like the genomes of apes. For more about this see "Human Genomic Organization and Introns" on our website (www.faithaliveresources.org/origins).

This scenario raises the same theological problems found as the Appearance of Age Interpretation discussed in chapter 5. Just as there are theological problems with the idea that God created the earth a few thousand years ago but created it to *appear* billions of years old, there are theological problems with the idea that God specially created the first humans without using common ancestry but made our DNA and the fossil record *appear* as though we share a common ancestor with animals.

Adam and Eve as Recent Representatives

In this scenario God created humans around 150,000 years ago, either using evolutionary creation or using miraculous modification. (In either case this scenario includes common ancestry with animals.) Then, about 10,000 years ago, God *selected* a pair of humans, Adam and Eve, to represent all of humanity. When Adam and Eve, humanity's representatives, chose to sin, their sinful status was applied to all human beings. The descendants of Adam and Eve mixed with the descendants of other humans alive at the time, and humans today have other ancestors in addition to Adam and Eve.

This scenario does not contradict the fossil record, archeological evidence, or genetic evidence we see today. These lines of scientific evidence would reflect how God created the bulk of humanity. Scientific evidence is not capable of confirming or denying the existence of a *particular* pair, Adam and Eve, because so many other humans were alive at the time.

This scenario was actually introduced before modern science because it resolves issues like Cain's wife and the large population indicated in Genesis 4. Also, the Genesis 1:26-28 account of the creation of humanity does not specifically refer to an initial pair of humans; it simply refers to humanity as *them*. Proponents of this view suggest that Genesis 1 refers to the initial creation of all humanity and that the following chapters in Genesis refer to a later time when God specially created or selected Adam and Eve and revealed himself to them.

This scenario introduces several new and difficult theological questions. Much of Genesis 2 and the following chapters refer to Adam and Eve as though all humans had descended from them. The original author seemed to assume that Adam and Eve were the very first people, and this is probably how the original audience heard it.

This scenario is compatible with traditional views of humans as God's imagebearers. God could have given humans their mental and social abilities through evolutionary creation or miraculous modification. This scenario leaves open the question of whether or not humans living *before* Adam and Eve had been commissioned to be God's stewards of the earth. But once God had revealed himself to Adam and Eve, they had a relationship with God and a commission to act as stewards of the earth. Since Adam and Eve were representatives, that commission extended to the rest of humanity.

In a similar way this scenario is compatible with traditional views on the relationship between body and soul. Although humans living before Adam and Eve would have had souls, this scenario leaves open the question of what sort of relationship, if any, they had with God.

In this scenario humans living before Adam and Eve clearly were naturally mortal before Adam and Eve sinned. This is most easily compatible with the view that Adam and Eve were naturally mortal and that the Fall resulted only in spiritual death. It could also be compatible with the view that Adam and Eve were naturally mortal but *potentially* immortal. Had they not sinned,

God could have provided them, and perhaps the rest of humanity, a miraculous transformation to immortality (the tree of life), either immediately or at some future point. This scenario is most difficult to reconcile with the view that Adam and Eve were *naturally* immortal before they sinned. For this to be the case God would have had to first miraculously transform Adam and Eve's mortal bodies into immortal bodies—including dramatic biological changes—and then transform them back into mortal humans after they sinned.

In this scenario Adam and Eve were in an actual state of original righteousness after God selected them and revealed himself to them. They fell into a state of sin through a disobedient act. How was that state of sin transmitted to the rest of humanity? Not all of the transmission methods described in chapter 11 make sense in this scenario. If original sin is transmitted *biologically* through birth, then this scenario does not work. Any biological change in Adam and Eve would only have affected their descendants, not all the other humans then alive. If original sin is transmitted *socially*, the scenario could work, since Adam and Eve interacted with other humans alive at the same time. If original sin is transmitted *spiritually* through birth, the representative scenario does not work. However, this scenario does work if original sin was transmitted *spiritually* by a representative process. The spiritual *state of sin* could have been applied to the rest of humanity because Adam and Eve acted as their representatives.

Some theologians argue that our legal status before God *does* function this way—through our representatives and not by birth. Romans 5:17-19 emphasizes the representative role of Adam and sees no theological problem with many people being declared sinful through the actions of a representative. Similarly, Christ acts as our representative—many are declared righteous through him, even though we haven't earned it.

Another theological issue with this scenario is the moral responsibility of the humans who lived and died in the thousands of years before Adam and Eve. Would they have known whether their behavior was right or wrong? They probably had the intelligence to understand the difference between truth and lies and between actions that helped or harmed others, but they had not yet been given a revelation of God's standards. Were these humans who lived before Adam and Eve in an ambiguous spiritual state, declared neither righteous nor sinful until Adam and Eve acted? Some proponents of this view try to resolve this difficulty by saying that *sin* is defined only in terms of our relationship with God and not in terms of its effect on other human beings. They argue that the humans who had not received a revelation from God were not held accountable for their behavior until after Adam and Eve's disobedience. But the Bible and Christian theology describe sin both in terms of our relationship to God and in terms of how it affects others and ourselves. So the question remains: did humans live and die for tens of thousands of years before Adam and Eve without sinning, without being held responsible to God for their actions?

Proponents of this view point to parts of Romans 5:12-19 to explain the status of humanity before Adam and Eve. In verses 13-14 Paul writes,

> To be sure, sin was in the world before the law was given, but sin is not charged against anyone's account where there is no law. Nevertheless, death reigned from the time of Adam to the time of Moses, even over those who did not sin by breaking a command, as did Adam, who is a pattern of the one to come.

Paul's argument is for the people living between the time of Adam and the time of Moses. Proponents of this scenario assert that what Paul says could be extended to people living before Adam. If sin is not taken into account when there is no law, then perhaps they had a different moral status before God than did the people who lived after God revealed himself to Adam and Eve.

A related theological issue is the spiritual status of humans who lived during and after the time of Adam and Eve but who lived too far away to have communicated with Adam and Eve or their descendants. From a human perspective it doesn't seem fair that God would declare all of these people sinful without their knowing about it or having a chance to affect Adam and Eve's choice. On the other hand the issue of fairness is not unique to the question of Adam and Eve. All people today are born under sin, without a chance to participate in Adam and Eve's choice. Just as people at the time of Adam and Eve hadn't yet heard, many people around the world today haven't yet heard the gospel. The theological answers applied to these questions today could also be applied to the people living around the time of Adam and Eve.

Adam and Eve as a Pair of Ancient Ancestors

In this scenario God used the natural mechanisms of evolution to create *hominids*. About 150,000 years ago God selected a pair of hominids and miraculously modified them into the first humans, Adam and Eve. This was certainly a spiritual transformation, and it included enough mental and physical changes that they became a separate species. All modern humans are descended from this pair and have inherited their sinful status. Humans are not descended from other hominids.

Like the recent ancestors scenario, this scenario is easy to reconcile with centuries of Christian theology about humans being created in the image of God, our fall into sin, and human mortality before the Fall. Since this scenario involves some miraculous action in the creation of Adam and Eve, this scenario is also compatible with all of the traditional views about the soul. It allows a reading of Genesis 2 and 3 with Adam and Eve as the first pair of humans. It does add one new interpretative difficulty: the very long period (around 150,000 years) between the Adam of Genesis 2 and 3 and the culture described in Genesis 4. The agriculture, music, metalwork, and cities described in Genesis 4:17-22 are not seen in archeological evidence until about 8,000 years ago. The genealogies in Genesis may have skipped some generations, but it seems unlikely that those skips could cover a gap of 142,000 years; such a gap must have included significant changes in culture, language, and so on.

The ancient ancestors scenario is more consistent with the fossil and genetic evidence than the recent ancestors scenario. In 150,000 years Adam and Eve's descendants would have had time to spread around the globe to every continent. Since God used only small miraculous genetic changes in creating the first true humans from hominids, this scenario is consistent with the large number of genetic similarities between humans and animals. The miraculous transformations made by God could not have been too dramatic: scientists have sequenced the entire genomes of both humans and chimps, and the differences between the two species are consistent with an ordinary rate of mutation since the time of the common ancestor. But by proposing a small number of carefully chosen physical and genetic transformations, this scenario remains consistent with the genetic evidence, while still giving a miraculous beginning to the human race, which is one way of clearly establishing our spiritual status before God.

But this scenario is difficult to reconcile with the genetic diversity seen in humans today. If God had started by transforming a pair of hominids into humans, and we are descended from that single pair—even if that happened over 150,000 years ago—then we shouldn't find so many different *alleles* in the human population today. God would have had to miraculously increase the number of mutations in various genes over several generations to produce the genetic diversity we see in the human population today. In this scenario God created us from a single pair but made it *appear* as though humanity had descended from more than a single pair by miraculously increasing the genetic diversity of their descendants. As with the recent ancestors view, this view also raises the theological difficulties found in the Appearance of Age Interpretation.

Variation: Group of Ancient Ancestors

A variation of the ancient *pair* scenario is the ancient *group* of ancestors scenario. In this variation somewhere around the time of the *population bottleneck* (see ch. 11, p. 236) in human history, God revealed himself in a single event to a large group of humans, and they fell into sin. The Adam and Eve story in Genesis 2-3 is symbolic of what happened to this larger group. We are all descendants of this group.

Proponents of this variation argue that this *group*, even if it is not a single pair, at least constitutes the sole ancestors of humanity. This avoids the theological difficulties raised by the representative scenarios in which humanity has ancestors besides those who received the first revelation.

This view is consistent with most of the fossil and genetic evidence, including some of the genetic diversity of humans today. God could have used the mechanisms of evolution to bring about this group of humans and then spiritually transformed them via his revelation, or he could have performed some degree of physical and genetic transformation on this group at the same time.

But this variation doesn't eliminate all difficulties. The genetic evidence for the population bottleneck in human ancestors indicates that the bottleneck was probably 10,000 individuals or more, a far larger population than could have been living together in one place at a time when humans lived in much smaller hunter/gatherer bands. If God had revealed himself to this many individuals at once, this would have had to happen to many different small groups at the same time. In terms of genetics the size of this group of ancestors could in principle be reduced to about 75 individuals, a small enough number that God's revelation could have happened to a single group. This number is just large enough to account for the genetic diversity in the human population today without the need for God to have performed miracles in subsequent generations. This number assumes that when God transformed this group from hominids into the first true humans he created all of the necessary genetic diversity in it and then ensured that this group had enough descendants to maintain that level of genetic diversity into later generations. This variation might still have some Appearance of Age theological difficulties, but they are smaller than the difficulties caused by supposing that all humans descended from a single pair.

Adam and Eve as a Group of Ancient Representatives

In this scenario God created humans around 150,000 years ago through evolutionary creation. God revealed himself to a group of humans rather than to a single pair. Other humans besides this group were alive at the same time, but this group was chosen by God to *represent* the rest of humanity alive at the time. These representatives chose to sin, and their sinful status was applied to the rest of humanity. Their descendants mixed with the descendants of other humans alive at the time.

This scenario matches the evidence from God's revelation in nature without difficulty. It does not contradict fossil evidence for the spread of humanity around the globe nor the genetic evidence for common ancestry with animals. It is also consistent with the genetic diversity seen in humans today.

This scenario answers most theological questions the same way as the recent representatives scenario. It has similar theological strengths and weaknesses. In particular, it faces the same theological challenges regarding the spiritual status of other humans living at the time of God's revelation who were not part of this representative group.

Biblically, this scenario would require an *allegorical* rather than a literal interpretation of Genesis 2-3 with regard to Adam and Eve in the garden. Even before modern science some Christians preferred an allegorical interpretation because many parts of the story, such as the talking snake and the tree of life, sound clearly allegorical. They argue that an allegorical interpretation does not deny the historical nature of these chapters. In an allegory the characters and plot are references to real people and real historical events, but the historical details are not recorded or are replaced with a more familiar context. In this view the story tells us the essential information about what happened regarding God's revelation to humanity, humanity's temptation, and their choice to disobey God's will, but the story is not intended to give the details of how it actually occurred.

This scenario allows for the fall of humanity into sin to be a single, distinct historical event, in contrast to the symbolic scenario we will describe below. Supporters argue that *some* historic event must have been the first personal revelation of God to human beings. That event would certainly have changed their relationship to God and would have given them moral responsibility toward him. When humans chose to act contrary to God's will, sin entered the world. Even if we don't know exactly how and when this happened, humanity's moral status changed at a particular moment in history.

Supporters of this scenario argue that it can preserve traditional theological teaching about original sin yet still be consistent with the scientific evidence for the evolutionary creation of humans. God could have made humans from hominids using the natural processes of mutation and natural selection, under his providential control, without any miraculous addition of mental or physical abilities. The key to the new spiritual status of humanity was not the miraculous creation of a new species, or even a miraculous physical transformation, but the first *special revelation* from God. With that revelation, humanity began to relate to God personally, and with that revelation they experienced their first temptation to rebel against God.

Adam and Eve as Merely Symbolic

In this scenario God created humans around 150,000 years ago through evolutionary creation. The human race grew into its current moral and spiritual status over time. Our sinful nature developed at the same time as our moral sense, our ability to reason, and our ability to communicate. No particular single event marks the time when all humans fell into sin; rather, many events occurred in which various human individuals and groups rebelled against God.

Like the previous scenario this is consistent with all current scientific evidence. Because it involves evolutionary creation of humans, including gradual development of human mental and social abilities, this scenario is usually associated with the view that our souls are a combination of our mental capacities plus our spiritual relationship with God. It is possible, however, to combine this scenario with views in which the soul is miraculously created.

This scenario is compatible with traditional views of humans as God's imagebearers; in this case humanity's status as God's image-bearers developed gradually over time. Our relationship with God and our commission to act as his caretakers of the earth would have happened through several revelations from God over time.

In this scenario humans would have been naturally mortal before the Fall. It is compatible with the view that the Fall caused only spiritual death and could also be compatible with the view that humans were naturally mortal but had the potential for immortality—had early humans not chosen to sin, God could have provided a means to immortality at some point.

This scenario requires that Genesis 2 and 3 be interpreted as a symbolic representation of truth. The Adam and Eve story teaches that humanity is made in God's image and that humanity rebelled against God's revealed wishes. It does not refer to a single event; rebellion could have happened many times, in many geographical locations, and in many human groups. This interpretation of Genesis 3 causes several theological difficulties because much of our understanding of the Bible and the gospel is tied to the Fall as a particular historical event.

This scenario is compatible with the views of *social* and *biological* transmission of original sin, but it doesn't work as well with the view of transmission of original sin by *spiritual status*. Rather than a single event in which humanity's fellowship with God is broken, it advocates a string of several events. Humanity's spiritual status would have been ambiguous until sin had spread throughout all of humanity. Reconciling the idea of a symbolic Adam and Eve with centuries of theological writing on the transmission of original sin as a one-time change in spiritual status would take considerable theological work.

A significant theological problem in this scenario is that the creation and the fall of humans, while theologically distinct ideas, both happen simultaneously as gradual processes. The Fall does not occur after the creation of humans but throughout the development of the human species. Humanity's moral and intellectual abilities and their moral and spiritual failings grew together. With each new ability, humans made sinful choices. As God gave humans greater moral and intellectual abilities, God's *special* revelation of his will also increased. These revelatory events could have happened through direct revelation, a prophetic voice, or the Spirit moving the conscience but could not have happened as a *single* revelatory event for all human beings at the same time. Proponents of this scenario compare this to the situation of human infants today as they simultaneously develop their moral, mental, and physical capabilities over several years. Traditional theology says that infants today are born in a *state of sin*, but we don't usually think of newborn infants as committing sinful *acts*. But at some time in early childhood, before their mental and moral capabilities are fully formed, we start to think of some of their actions as clearly sinful.

In this scenario the *state of original righteousness* is merely a potential state, not an actual state. It is a state that humanity *might* have achieved had it responded obediently to God's revelations. Instead, humans repeatedly chose to disobey God and gradually fell into sin. Because their mental capabilities were developing at the same time as their tendency to sin, there was no moment in history when humans had a full set of human capabilities and were also morally righteous. Thus, humanity never was in a fully human state of original righteousness. This view is difficult to reconcile with centuries of Christian theological writing that links the doctrine of original sin with belief in an *actual* state of original righteousness. Those who view Adam and Eve as symbolic argue that this link is in the writings of the church fathers, not because the beliefs are *necessarily* linked but because the church fathers had always assumed that Adam and Eve were literal ancestors.

The symbolic view of Adam and Eve is also difficult to reconcile with Romans 5:12 ("Sin entered the world through one man") and the following verses. These verses also say that humanity was not always sinful and that sin came through a choice by the first humans. Supporters of this scenario argue that if we use the hermeneutical principle of putting ourselves in the mind-set of the original author and audience in order to determine its intended message, then this passage is not intended to *teach* that Adam and Eve were real persons; rather it intends to teach that salvation can come to many through Christ. As Denis Lamoureux, a professor of science and religion, puts it,

> The context and intention of these passages are not debating the historicity of Adam. They focus on the reality of sin and the fact that Jesus frees men and women from their sinfulness and offers them resurrection from death and eternal life. Moreover, it must be remembered that Paul employs the science-of-the-day in his inspired letters. He uses the 3-tier universe in one of the most significant passages in the New Testament—the kenosis of Jesus (Phil. 2:5-11). Consistency demands that since this apostle holds an ancient understanding of the structure of the universe, then he undoubtedly accepted an ancient view of human origins—*de novo* creation.
>
> —*Evolutionary Creation: A Christian Approach to Evolution,* p. 274.

The theological difficulties posed by the symbolic scenario make it more challenging to reconcile with the teachings on original sin by St. Augustine, as well as by other theologians in the Roman Catholic, Lutheran, Calvinist, and many evangelical traditions, although there are some theologians in those traditions who favor this scenario. Some other Christian traditions, notably the Eastern Orthodox, find it less difficult to connect their teachings on original sin with this scenario. (They have traditionally viewed Adam and Eve as childlike.) Because this scenario is more difficult to reconcile with traditional Western theology and Scriptural interpretation regarding humanity's creation and fall into sin, theologians would have to do considerable work before this view would be widely accepted.

Variations on These Scenarios

Having read these scenarios, you can probably think of variations. For example, maybe God miraculously created Adam and Eve from dust but made their bodies and their genes consistent with hominids alive at the time (the *ancestor* views) or consistent with other humans living at the time (the *representative* views). This would have some of the theological problems of the Appearance of Age Interpretation, but it would be more consistent with some interpretations of Genesis 2.

Or perhaps there was a *pair* of ancient representatives rather than a *group* of ancient representatives. This scenario would still be consistent with the scientific evidence but also allows a less allegorical interpretation of Genesis 2-3.

THEOLOGICAL AGREEMENT AND DISAGREEMENT AMONG THESE SCENARIOS

Christians in Agreement

While these five scenarios have serious differences with each other, they also agree with each other on some key points. In all of these scenarios human beings are uniquely God's imagebearers,

▶ gifted by God with certain abilities.
▶ invited by God into a personal relationship.
▶ commissioned by God to be stewards of this earth.

All of these scenarios can be compatible with Christian beliefs about the body and the soul. All five scenarios can also be compatible, some more easily than others, with at least two views on human mortality before the Fall:

▶ the Fall caused only spiritual death.
▶ the Fall caused both spiritual and physical death, but humans were only potentially immortal by God's grace.

All of these scenarios agree about the *situation* of original sin. They agree that

▶ humans today are sinful and in a broken relationship with God.

▶ humans cannot achieve righteousness by their own action.

▶ we can only be redeemed through the work of Christ.

Christians in Disagreement
These five scenarios primarily disagree about the following questions:

▶ How and when did humanity fall into that sinful state?

▶ Was the first sin committed by our ancestors or by our representatives?

▶ What was the spiritual status of any humans living before the first sin?

To some Christians these are vital questions, while to others they are secondary. Some argue that a clear historical first sin, committed by Adam and Eve as our ancestors, is essential to our understanding of Christian theology. Others agree with Lutheran theologian George Murphy:

The Christian claim is that a savior is needed because all people are sinners. It is that simple. *Why* all people are sinners is an important question, but an answer to it is not required in order to recognize the need for salvation. None of the gospels uses the story in Genesis 3 to speak of Christ's significance. In Romans, Paul develops an indictment of the human race as sinful and then presents Christ as God's solution to this problem in chapters 1-3 before mentioning Adam's sin in chapter 5.

—"Roads to Paradise and Perdition: Christ, Evolution, and Original Sin," *Perspectives on Science and Christian Faith,* June 2006.

Not Satisfied with Any of These Scenarios?

Neither are we! All of the Adam and Eve scenarios discussed in this chapter seem to have significant scientific or theological challenges or both. While you likely have disagreements with one or more of the views presented, it's worth considering each one at least briefly. Don't simply dismiss them for sounding "wrong," but figure out exactly which arguments you disagree with and why. If you get a chance, discuss these ideas with other Christians. Remember that proponents of each view can be working in good faith to reconcile God's revelations in Scripture and in nature and to maintain certain central theological beliefs.

The Bible is useful for teaching, training in righteousness, and equipping the people of God (2 Tim. 3:16-17), but it does *not* answer every question we could ask or imagine. That means that we are left with ambiguity about the details. This ambiguity leads to disagreements among Christians about what Scripture is really teaching. Because these issues *are* complicated, we've chosen to explain a variety of options and to explore the theological and scientific issues at stake in each option so that you can make your own decision.

As we stated in chapter 4 in our discussion of the two-books diagram of nature and Scripture, when a conflict arises our response should be to examine *both* the science *and* the biblical interpretation more carefully. Conflict resolution also requires that we listen to a range of views. When Christians disagree strongly on a particular issue, they can still agree that proper understanding of both nature and Scripture is essential to the solution and that the conflict is at the human level.

The church has work to do in this area. This work is not something to fear but rather is a part of the church's calling. In the words of Christian geologist Davis Young, the church should

encourage Christian theologians, anthropologists, archaeologists, and paleontologists to collaborate in honest, forthright assessment of the available evidence and to develop a viable position that preserves the biblical doctrines of man, sin, and salvation. We can rejoice whenever Christian scholars are together driven to a closer scrutiny and deeper appreciation of the Word and works of God.

—"The Antiquity and the Unity of the Human Race Revisited," *Christian Scholar's Review,* May 1995.

QUESTIONS FOR REFLECTION AND DISCUSSION

1. What do you think of the scenarios describing Adam and Eve as recent or ancient *ancestors*? How would you deal with the scientific evidence, especially genetic diversity in the human population, that doesn't fit any of these scenarios?
2. What do you think of the scenarios describing Adam and Eve as recent or ancient *representatives*? In these scenarios, what do you think was the spiritual status of humans who lived before Adam and Eve? Of those who lived after Adam and Eve but never had a chance to interact with them or their descendants?
3. What do you think of the scenario that describes Adam and Eve as merely symbolic? Is it possible to reconcile this scenario with what Scripture teaches about the creation and fall of humanity? With what theologians have written over the centuries regarding the creation and fall of humanity? Why or why not?
4. Near the end of the chapter several paragraphs discuss ways in which these different scenarios agree with each other. What other areas of agreement can you think of?

ADDITIONAL RESOURCES

Collins, C. John. "Adam and Eve as Historical People, and Why It Matters," *Perspectives on Science and Christian Faith,* 62:145, September 2010.

Collins, Robin. "Evolution and Original Sin," *Perspectives on an Evolving Creation.* Keith B. Miller, ed. Grand Rapids, Mich.: Wm. B. Eerdmans, 2003.

Harlow, Daniel C. "After Adam: Reading Genesis in an Age of Evolutionary Science," *Perspectives on Science and Christian Faith,* 62:179, September 2010.

Hurd, James. "Hominids in the Garden," *Perspectives on an Evolving Creation.* Keith B. Miller, ed. Grand Rapids, Mich.: Wm. B. Eerdmans, 2003.

Murphy, George. "Roads to Paradise and Perdition: Christ, Evolution, and Original Sin," *Perspectives on Science and Christian Faith,* 58:109, June 2006.

Young, Davis. "The Antiquity and the Unity of the Human Race Revisited," *Christian Scholar's Review* XXIV:4, May 1995. Available at www.asa3.org/asa/resources/CSRYoung.html.

Arguments for the Recent Representatives scenario:

Alexander, Denis. "How Does a BioLogos Model Need to Address the Theological Issues Associated with an Adam Who Was Not the Sole Genetic Progenitor of Humankind?" BioLogos white paper, biologos.org/uploads/projects/alexander_white_paper.pdf.

Berry, R. J. and Jeeves, M. "The nature of human nature," *Science & Christian Belief,* 20:3-47, 2008.

Arguments for the Group of Ancient Ancestors scenario:

Day, A. J. "Adam, anthropology and the Genesis record—taking Genesis seriously in the light of contemporary science," *Science & Christian Belief,* 10:115-43, 1998.

Arguments for the Symbolic scenario:

Lamoureux, Denis O. *Evolutionary Creation: A Christian Approach to Evolution.* Eugene, Ore.: Wipf and Stock Publishers, 2008.

Schneider, John R. "Recent Genetic Science and Christian Theology on Human Origins: An 'Aesthetic Supralapsarianism,'" *Perspectives on Science and Christian Faith.* 62:196, September 2010.

CHAPTER 13

BUT WHAT ABOUT . . . ?

After reading the first twelve chapters of this book, you probably have a lot of questions. We've raised numerous issues and left many questions unanswered. In earlier chapters we

▶ described several interpretations of Genesis 1.

▶ summarized scientific evidence that the earth is billions of years old and that the theory of evolution is an accurate description of the history of life.

▶ laid out a spectrum of views Christians hold in an attempt to bring science and biblical interpretation together.

▶ discussed a spectrum of views Christians hold on the origin of humans.

Among those many views none is free of theological or scientific challenges. All raise difficult questions. To help you think through some of the key issues, we've put together a list of 26 questions Christians often ask about origins. You'll find the complete list of questions in the section "Questions for Reflection and Discussion" (p. 284) at the end of this chapter. In addition, we discuss all of these questions on our website (www.faithalive resources.org/origins). Click on "Questions Christians Ask . . . ." Many of the questions have been addressed in previous chapters, so we will discuss just six of them in this chapter:

- ▶ Since the Bible tells us how God made the world, why do we need to listen to science?
- ▶ Shouldn't there be some sort of proof in nature that God created it?
- ▶ Would humans be less significant if God had created us through common ancestry with animals, rather than through special miracles?
- ▶ With all this disagreement in the church, what should I believe?
- ▶ What should I teach my children?
- ▶ How do I deal with disagreements about origins with my family and church members?

Since the Bible tells us how God made the world, why do we need to listen to science?

God both created nature and inspired Scripture. As we discussed in chapter 2, both are revelations from God that have something to teach us. Many Bible passages, such as Psalm 19, point to God's revelation in the natural world. Because they are both revelations from God, nature and Scripture cannot conflict with each other. Conflict comes at the level of human interpretation of one or both revelations. If someone says "The Bible trumps science," they are really saying that their human *interpretation* of the Bible trumps a scientific interpretation of nature.

Also, the Galileo incident (ch. 4) shows us that the Holy Spirit can sometimes use discoveries of science to prompt us to reexamine our interpretation of Scripture, leading us ultimately to a better understanding of Scripture. We should not neglect this means by which God can teach us new things.

Shouldn't there be some sort of proof in nature that God created it?

It's understandable that Christians would want to see proof of God in nature. This, in part, motivates some Christians to try to find scientific proof that the earth is young or that the theory of evolution is false.

If he wanted to, God certainly could plant evidence for miracles all over nature. God could also perform obvious miracles for every human being who ever lived, but he chooses not to do that. The Bible teaches that God did perform dramatic miracles and that sometimes people responded to those miracles with faith. But at other times these miracles did not result in lasting faith in God. When the Israelites came to Mount Sinai, they saw the whole mountain covered in lightning and billowing smoke; in the midst of the thunder they heard the voice of God (Ex. 19). The experience was so fearsome that they begged Moses to be God's spokesman and committed themselves to doing whatever God said. But only days later they made a golden calf to worship. When the Israelites went to Mount Carmel in the time of the prophet Elijah, they witnessed God miraculously sending fire on Elijah's sacrifice in a contest with the prophets of Baal (1 Kings 18). Did this cause everyone present to turn from Baal to the Lord? Days later Elijah had to flee for his life. When the Pharisees saw Jesus perform many miracles, they understood that no one can do miracles apart from God (John 3:2; 9:16). Despite these miracles nearly all of the Pharisees rejected Jesus, and many called for his crucifixion.

These and other Scripture passages reveal that even when faced with obvious proof of God's existence some people choose not to respond in faith (see also Jesus' parable of the rich man and Lazarus, Luke 16:19-31).

In Romans 1:20 we read, "Since the creation of the world God's invisible qualities—his eternal power and divine nature—have been clearly seen, being understood from what has been made, so that people are without excuse." One way to interpret this passage is to say that nature must provide proofs of God's existence in the form of something that science cannot explain. Some proponents of Intelligent Design theory point to the genetic complexity of living organisms as one such proof. Other Christians say that it would be strange for God to bury such a proof of his existence in ways that only modern genetic science could detect.

Does Romans 1:20 actually teach that nature provides these sorts of proofs of God? Consider the context of verses 21-23:

For although they knew God, they neither glorified him as God nor gave thanks to him, but their thinking became futile and their foolish hearts were darkened. Although they claimed to be wise, they became fools and exchanged the glory of the immortal God for images made to look like mortal human beings and birds and animals and reptiles.

These verses show that Paul was thinking about the pagan idolatry of his time. People steeped in this idolatry took one created thing (like the sun or the moon or the sea) and called it a god, or they took one aspect of creation (like fertility or death) and worshiped it. Instead of worshiping the Creator, ancient pagans took one *part* of the creation and looked to it for hope and meaning.

In this sense modern scientific atheism is somewhat like pagan idolatry. Such atheists take one aspect of nature—the regular functioning of natural laws—and turn it into a god. It becomes the foundation of all their hopes and beliefs about the world, as in this quote by chemist P. W. Atkins:

Scientists, with their implicit trust in reductionism, are privileged to be at the summit of knowledge and to see further into truth than any of their contemporaries.... They are the beacons of rationality, lighting the trail for those who wish to use that most powerful and precious of devices, the human brain.... Science, with its currently successful pursuit of universal competence through the identification of the minimal, the supreme delight of the intellect, should be acknowledged king.

—P. W. Atkins. "The Limitless Power of Science,"
*Nature's Imagination: The Frontiers
of Scientific Vision.* J. Cornwell, ed.; 1995.

Ancient pagans and modern atheists alike have rejected the true God revealed in the regular functioning of natural laws and have turned a created thing into an idol. The answer to the ancient pagans was not to claim that the sun or the sea or fertility didn't exist but to put these things in their proper place as aspects of God's creation. Considering today's context, Romans 1:20 teaches that the answer to modern atheists is not to deny the regularity of natural laws or to look for miraculous breaks in them but to put natural laws in their proper place as God's creations. Of course, God certainly does use miracles at times to reveal himself. But Romans 1:20 does not seem to teach that nature *must* contain miraculous proofs of God.

Would humans be less significant if God had created us through common ancestry with animals rather than through special miracles?

The idea of human evolution raises concerns about human significance. If we evolved from animals, are we nothing more than animals? Even if humans share a common ancestry with apes and other animals, our line of descent diverged from that of other animals at some point. Something different happened in our line of descent that did not happen to apes or other animals, something that makes us unique among life-forms on earth.

Our significance, however, lies not primarily in our biological uniqueness but in how God chooses to relate to us. In Genesis 1-2 God did more than create our bodies. He chose to reveal himself to human beings, establishing a relationship with us beyond the relationship he has with animals. God blessed humanity and declared it very good. He gave humanity a commission he did not give to other animals: to name the creatures and to exercise stewardly dominion over God's creation.

While God continues to care and provide for animals, throughout the Old Testament we see God doing dramatically more than that for human beings. God revealed himself personally to his people through word and action, establishing covenants with them and answering their prayers. Although humanity is small compared to the size of the universe, the vastness of the created world is not meant to belittle us but to proclaim the vastness of God's promises (Gen. 15:5) and of his love (Ps. 104:11-12). Our significance is based on our standing in God's eyes, not on our physical size or uniqueness.

Beyond all this God chose to become incarnate as a human being, to take on our very form. Jesus Christ humbled himself and took on a human body. That act alone raises the significance of humanity. "God demonstrates his own love for us in this: While we were still sinners, Christ died for us" (Rom. 5:8).

How can our significance depend on whether we share a common ancestor with apes? God has done things for humanity that go far beyond his relationship with any other species. Our significance and worth in his eyes are undeniable, regardless of what we discover about *how* God created us.

With all this disagreement in the church, what should I believe?

That's the biggest question, and of course we can't answer it for you. We understand that this issue can be overwhelming. After reading a book full of evidence and arguments, it's normal to feel a little uncertain what to think. If this were an easy issue, Christians would already agree!

Look back at the spectrum of young-earth creation, progressive creation, and evolutionary creation views in chapter 8 (pp. 187-188) and in the Appendix. *Every* position raises theological challenges, and it takes time to ponder them. Don't feel as though you have to work out what you believe in a day or a month. For us and others we've known, it can take months or years of reading, pondering, and conversation to accept a new position.

Questions about how and when God created the earth are important, but they are not *essential* to our salvation. It is not essential for every Christian to come to a firm conclusion on these issues. It *is* important, however, that leaders be well informed. If you are a pastor, theologian, or scientist, if you speak publicly on this issue, or if you teach children about science, it is important to understand the various positions. When you speak from a position of authority people take your words seriously, so it is important that you avoid mistakes and misunderstandings. It's best not to promote one particular view without understanding and acknowledging the other views held by Christians. Having read this book, you're well on your way! If you'd like to learn more, check out the additional resources suggested at the end of each chapter and the articles on our website (www.faithalive-sources.org/origins).

Regardless of what you decide about origins, keep these things in mind as you discuss these issues with others:

▶ Fight against the worldviews of evolution*ism* and naturalism. Challenge claims that a scientific understanding of the Big Bang or evolution somehow disproves God. Whether or not the Big Bang and evolution happened, God is the sovereign Creator. Science can't prove or disprove that.

▶ Remember that all truth is God's truth. Even when an idea is promoted by an atheist or by someone you dislike, it is not automatically false. Be willing to consider true arguments from any source, and know that God owns all truth.

▶ Avoid adding to the gospel. Keep the gospel centered on the work of Christ and our need for grace, independent of views on origins. When non-Christians hear Christians make blanket scientific statements on origins, they get the impression that they'd have to change their scientific views in order to become a Christian.

What should I teach my children?

For young children, start with Bible stories. The Genesis stories are short and emphasize essential truths, so they are excellent for teaching children. Don't try to teach them any science or theology beyond what they are ready to learn.

Children can be taught a sense of wonder and joy when learning about the natural world. It comes naturally to them, but adults can nurture it and expand it. And when they express that wonder and joy, remind them to thank God for making all of it. When a child asks "Did God make the dinosaurs," you can simply and truthfully answer "Yes," without getting into the details about how God did it. You can tell children that the Bible tells us that God made everything and that God lets us study the world to figure how it works.

In elementary school children start to learn more Old Testament stories, including the idolatry of the nations surrounding Israel. This is a good opportunity to explain ancient Near Eastern cosmology and how Genesis 1 is a radical response to the paganism and polytheism of the surrounding cultures (ch. 6).

In middle school and high school teens are encountering the science of evolution in school. Developmentally, they are better able to appreciate that not all questions have black and white answers. This is a good time to introduce them to multiple views on origins, explaining where Christians agree and disagree (see the end of ch. 1 for some educational resources). It's important that teenagers learn about the full range of views that Christians hold on origins. When they are taught only one particular perspective as the "right" view, they tend to consider it essential to Christianity. This can lead to a crisis of faith later on, a feeling that they will lose their entire Christian faith if they change their mind about origins. Even if you (or they) feel one view is best, let them know that there are other Christian views and discuss the options.

You can also encourage teens and young adults to consider a career in science. Christian scientists work alongside scientists of all worldviews and have opportunities to share the gospel with them. In addition, Christian scientists have the joyful task of exploring God's world! Encourage those considering a career in science to seek out role models and professionals in the community they can shadow. See the resource list at the end of this chapter for more on a career as a scientist and a Christian.

How do I deal with disagreements about origins with my family and church members?

The first few chapters of this book focused on topics on which Christians generally agree. In later chapters we explored issues of greater disagreement. Disagreements among Christians are not always a bad thing. If an issue is complex, it's unlikely that any one person has all the right answers. We can learn from each other. But in order to do that, we need to practice the virtues of humility and patience—as well as the habit of curiosity. We can model those in ourselves and encourage them in others.

Be humble. Pride can make anyone too stubborn to listen to new ideas or too quick to discard an old belief. Keep listening seriously to all sides, admit when you don't understand fully, and change your mind if you feel the arguments warrant it.

Respect and affirm the intelligence, motives, and faith commitment of Christians on all sides of the debate. Be slow to judge; give room for everyone to grow in their understanding of God's book of nature and God's book of Scripture. Don't accuse someone of holding a view because they lack knowledge or of changing their mind because they lack faith.

Nurture our unity in Christ. Remember what unites us as Christians. Cherish the central points of faith on which Christians agree.

While the latter half of this book has focused on areas in which Christians disagree, we want to close the discussion by returning to areas of agreement. Having the "right" view on every issue is less important than that the church lives and works and worships in unity. Our brothers and sisters in Christ who disagree with us about some things don't disagree with us about *everything*. In fact, we do agree about the most important things:

God created and sustains the universe. The natural world gives testimony to God's power, creativity, and faithfulness. All parts of this universe are God's creation and under God's control; none of them are divine powers in themselves. God created humans and gave them a special place as his imagebearers and caretakers of this world. Science and Christianity are not at war. In fact, scientifically studying God's creation is one way in which we can joyfully explore creation and fulfill our mandate to be caretakers. We have all sinned, individually and corporately. We have turned away from God; we have hurt each other, ourselves, and this world. God is a God of grace and love. And God has given us the best possible news: our Creator is also our Redeemer.

The issues of origins we've discussed in this book are by no means the only place where science and Christianity intersect. Consider, for example, stem cell advances, artificial intelligence, or global warming. For some thoughts and advice on dealing other science and faith issues, see "The Next Hot Issue" on our website (www.faithaliveresources/origins).

QUESTIONS FOR REFLECTION AND DISCUSSION

Regarding interpreting Scripture:

1. Since the Bible tells us how God made the world, why do we need to listen to science? (See our answer on p. 276 in this chapter.)
2. Haven't Christians always believed in a young earth and a six-day creation?
3. Is it ever appropriate to change one's interpretation of Scripture to match science?
4. Isn't a non-literal interpretation of Genesis 1 just a slippery slope to denying the resurrection?
5. If Genesis 1 should be understood literally, what is the "firmament" created on day two?
6. Why didn't God just tell us about the Big Bang and evolution in Genesis?
7. Is it better if we can make the events of Genesis 1 line up with what science says, or if the message of Genesis 1 is independent of what science says?

Regarding interpreting nature:

8. How strong is the evidence for an old earth? For evolution?
9. Are the Big Bang and evolution just beliefs promoted by atheists to get around God?

10. Are scientists biased against religion and against God?
11. How can scientists be sure about the Big Bang and evolution if no one was there to see them?
12. Is there any scientific evidence for a young earth?
13. Could God have created the earth recently and made it appear old?
14. Can the scientific evidence for an old earth also be made to fit a young-earth model?

Regarding the goodness and fall of creation:
15. An old earth would mean millions of years of animal pain and species extinction. Didn't God create the world perfect at the beginning?
16. How could God call creation good if it included destruction, pain, and extinction?
17. Does evolution reward selfishness?
18. Did death exist before the Fall?
19. Does a non-literal view of Adam and Eve deny important doctrines about original sin and salvation?

Regarding how God works in nature:
20. Why would God use such a long, slow process when he could have created everything instantaneously?
21. How could it all have happened by chance?
22. If evolution is true, doesn't God seem weak and uninvolved?
23. Does evolution imply that God doesn't do miracles?
24. Shouldn't there be some sort of proof in nature that God created it? (See our answer on pp. 276-279.)
25. Would humans be less significant if God had created us through common ancestry with animals rather than through special miracles? (See our answer on pp. 279-280.)
26. How do I worship God if God used slow, natural processes instead of miracles to create each animal and plant? (See our answer in ch. 14.)

Regarding how to live in unity while still disagreeing about the particulars:

27. With all this disagreement in the church, what should I believe? (See our answer on pp. 280-281.)
28. What should I teach my children? (See our answer on pp. 281-282.)
29. How do I deal with disagreements about origins with my family and church members? (See our answer on pp. 282-283.)

You'll find our thoughts on all of these questions on our website (www.faithaliveresource.org/origins). Click on "Questions Christians Ask..."

ADDITIONAL RESOURCES

More about the goodness and fall of creation:
Munday, John C. "Animal Pain: Beyond the Threshold?" *Perspectives on an Evolving Creation.* Keith B. Miller, ed. Grand Rapids, Mich.: Wm. B. Eerdmans, 2003.

Snoke, David. "Why Were Dangerous Animals Created?" *Perspectives on Science and Christian Faith,* Vol. 56, June 2004.

Yancey, Philip. *Where Is God When It Hurts?* Grand Rapids, Mich.: Zondervan, 1977, 1990, 2002.

More about a career as a Christian in science:
American Scientific Affiliation (www.asa3.org). This professional society for Christians in science has many useful resources from a range of viewpoints.

Bancewicz, Ruth. *Test of Faith: Spiritual Journeys of Scientists.* Wipf & Stock, 2010. Ten of today's scientists discuss their Christian faith.

Graves, Dan. *Scientists of Faith: 48 Biographies of Historic Scientists and Their Christian Faith.* Kregel Publications, 1996.

Hearn, Walt. *Being a Christian in Science.* Downer's Grove, Ill.: InterVarsity Press, 1997.

List of Christian scholarly societies (www.apu.edu/faith integration/resources/societies). This list includes several organizations of Christians in the sciences as well as other academic fields.

More on educating children and youth:
Brouwer, Sigmund. *Who Made the Moon? A Father Explores How Faith and Science Agree.* Thomas Nelson, 2008.

See educational resources listed at end of chapter 1.

CHAPTER 14

WONDER AND WORSHIP

I n the Bible the most frequent references to the natural world are in the context of worship, of praising God as the Creator. Yet in the church today heated debates over origins and other science issues can detract from worship and praise. Christians hear pastors or Christian radio stations saying that scientists are trying to disprove God, or that scientific advances will challenge their faith. These issues are important, and we've addressed them earlier in this book. But when the church's response to science is dominated by debate, it can leave the negative impression that science is all about controversy, challenges, and divisions among Christians.

Thus it is important for churches to make a special effort to praise God for what science has discovered about God's world. Worship provides balance to the Christian conversations about science by highlighting the important essentials of our faith and reminding us all of the Christian unity we share. More than that, worship and meditation provide space to ponder God's revelation in nature so that we may listen to it more attentively. What does God's creation teach us about God and his character?

Imagine living in ancient Israel and looking up at the night sky. On a clear, moonless night, without the light pollution of today's cities, the Milky Way would have been a sparkling band across the black sky. David wrote:

> The heavens declare the glory of God;
> the skies proclaim the work of his hands.
> Day after day they pour forth speech;
> night after night they display knowledge.
> There is no speech, they use no words;
> no sound is heard from them.
> Yet their voice goes out into all the earth,
> their words to the ends of the world.
>
> —Psalm 19:1-4.

The heavens proclaim God's glory for all people to hear. The message crosses barriers of language and culture to reach people of every tribe and nation. Indeed, everyone who sees the Milky Way feels a sense of awe. Yet those who know Jesus as their Savior can hear the message more clearly and understand it more deeply. Christians don't merely feel a vague sense of wonder at the natural world; we marvel at it as the handiwork of the God we personally know and love as our Father.

Notice how God's two revelations, nature and Scripture, work together. Nature alone doesn't teach us everything about God; our first and best source of information about God is always his special revelation in Scripture. On the other hand, nature reveals things about God in a way that dry words on a page cannot. We read of the psalmist's desire "to gaze on the beauty of the Lord and to seek him in his temple" (Ps. 27:4) and wonder what that beauty must look like. Then we gaze on the beauty of the Milky Way, sparkling across the sky, and expand our imagination of what God's beauty and glory might look like. Or consider a Scripture passage like Job 12:13: "To God belong wisdom and power." This can evoke images of mere earthly political or military power, but the display of power in a thunderstorm (recall the discussion of Ps. 29 in ch. 1) drives it home to our senses: we are blinded by the flash of lightning, shaken by the crack of thunder, and blown over by the wind and rain. God reveals his glory and beauty and power in nature in ways that words cannot express.

Calls to Worship from Modern Science

Modern science and technology have vastly expanded humanity's understanding of nature and thus have given the church many more reasons to praise the Creator. The Milky Way is glorious, but when the Hubble Space Telescope points at the Milky Way it finds incredible swaths of light and smoke across light years of space (see p. 159). A thunderstorm certainly displays God's power, but astronomers have found far more tremendous displays of power throughout the universe. In a supernova (ch. 3) a star dies in an explosion so incredibly powerful that it outshines a billion other stars, reminding us that God's power completely surpasses all earthly power. And the Big Bang (ch. 7) is far more than a light switch turning on at the beginning of the universe; God created light in a universe-wide explosion of space itself, a beginning of space, time, matter, and energy far more powerful than all the light of today's universe put together. The very vastness of space is a display of the vastness of God's love for us (Ps. 103:11-12, as discussed in ch. 7).

Sometimes, however, Christians focus on praising God for what science does not understand. Recall our warning in chapter 2 against "god of the gaps" arguments. It's tempting to say that "X is so amazing that scientists can't explain it, so God must have made it!" While God certainly made all of the things science can't explain, a problem comes into play when we focus our attention there. What happens when science develops an explanation for X? Will our praise go silent? A better approach is to also praise God for all of the amazing and wondrous things that science *can* explain, such as those described above. A scientific explanation does *not* replace God, nor should it lessen our praise. In fact, it can enhance our praise because it enhances our understanding of how God governs the natural world. In a limited way we have the privilege of "thinking God's thoughts after him."

Modern evolutionary biology also teaches us about God's character, including his creativity in making the wide variety of Galapagos finches through microevolution (ch. 9). It can be hard to think about worship and evolution in the same sentence if you've always heard the word *evolution* in a negative context. Here are thoughts from each of us on how we've come to see God's character while learning about the science of origins:

Deborah writes:

I grew up hearing a young-earth creation view, but as an adult I began to investigate origins for myself. Over a year or two of reading books and discussing the issues with other Christians, I decided that an old-earth view, with at least some evolution, was the best way to reconcile the evidence in nature and in Scripture. But after reaching that intellectual conclusion, it took another few years for me to repattern my worship habits to match it. For instance, what should I think about while singing hymns like the following?

All things bright and beautiful, all creatures great and
    small,
all things wise and wonderful—the Lord God made
    them all.

Each little flower that opens, each little bird that
    sings—
he made their glowing colors, he made their tiny
    wings.

The purple-headed mountain, the river running by,
the sunset, and the morning that brightens up the sky.

All things bright and beautiful, all creatures great and
    small,
all things wise and wonderful—the Lord God made
    them all.
                                    —Cecil F. Alexander, 1848.

Throughout my entire upbringing I sang hymns such as this while picturing God walking through the Garden of Eden with each bird flying out of his hand in a separate, special miracle, or picturing C. S. Lewis's Aslan singing the rocks and hills into being. But now I had decided that God had created these things through natural processes over millions of years. If God made birds and mountains using evolutionary biology and tectonic plate motion, what do we praise God *for*?

Over the years I have found many good answers to that question, some of which you've encountered throughout this book. One answer is to see God working in the natural, long-term processes. For example, now when I sing hymns about God creating mountains, I picture God using the flow of magma under the earth's crust to slam the Indian continental plate into the Asian plate, a very slow but incredibly mighty push to raise up the snowy heights of the Himalayas.

I now find myself praising God for the glory of the system in addition to each individual thing in that system. Not only did God make each individual mountain, but God carefully designed a whole system to form all of the mountains on earth. When I sing hymns about God creating flowers I think of the evolutionary mechanisms he designed to produce not just a few kinds of flowers but an extravagant abundance with every variation of size, shape, color, and scent. God designed a system that creates abundant beauty and also makes each flowering plant well suited to thrive in its environment.

Ultimately, the best answer is to worship God for the *who* and *why* of creation more than for the *how* and *when*. God is our sovereign, all-powerful Creator, and he declared all things good.

Loren writes:

> Although I was taught a Young-Earth Interpretation of
> Genesis in my grade school years, learning about the scien-
> tific evidence of an old universe and biological evolution did
> not diminish my sense of wonder or threaten my worship
> and prayer. My church prepared me for this in four impor-
> tant ways when I was in grade school. First, I was taught
> that a scientific explanation for how something works
> does not replace God, so we do not need to fear advances
> in science. Second, I was taught from the history of the
> church—from events like the Reformation, the Galileo inci-
> dent, and the abolition of slavery—that the Holy Spirit can
> sometimes correct the manner in which the church under-
> stands some part of Scripture. Third, I was taught that we
> can improve our understanding of Scripture by learning
> about the history, language, and culture of the original audi-
> ence and author. Fourth, and most importantly, I was taught
> that the foundation of my faith does not lie in how I inter-
> pret Genesis 1; rather, it is in the grace of God evident in the
> incarnation, death, and resurrection of Jesus Christ.

So when I learned in college about the evidence for an old
earth and an old universe, it didn't worry me much. It did,
however, prompt me to study what theologians had written about Genesis 1. And once I learned from these biblical scholars about the old-earth interpretations of Genesis
(discussed in ch. 6), I quickly felt free to enjoy the beauty
in the sciences of cosmology, geology, and evolutionary
biology without fear.

Now I eagerly read articles about how solar systems form
and about distant galaxies billions of light-years away. I
eagerly learn about how erosion and the motion of continental plates on earth, slowly moving over millions of years,
create a vast array of ecological niches—high mountains,
low foothills, plains, river deltas, lakes, sandy shores, shal-

low oceans, deep ocean trenches—each of which is home to a unique array of living organisms. Each time I learn more I'm filled with wonder and awe, as well as with a renewed desire to praise God.

Learning evolutionary biology also deepens my appreciation for what God has made. A few years ago I was walking on the high ridges in Rocky Mountain National Park. I saw tiny flowering plants that were adapted to life on those cold, windy ridges. I had seen similar plants at lower altitudes that were different species but closely related. I thought about how the processes of evolution, which God designed, allowed plants and animals to adapt to different environments over time. When a species of plant or animal lives in one ecological niche at the boundary of a second niche, the processes of mutation and reproductive success can, over time, result in some of them gradually spreading into the neighboring niche. Where once there was one species of daisy or shrub or squirrel or ant, after a while there become two such species, each adapted to its particular niche. God has created a marvelous mechanism for species to adapt to changing environments and to spread into new environments. And as they do, the process increases the beauty, diversity, and complexity of the natural world. Now I eagerly read articles about evolutionary biology. Each time I learn more I'm filled with wonder and awe, as well as with a desire to praise God.

Engaging Science in the Life of Your Congregation

We hope that when you set this book down you'll continue to ponder what God has revealed in the natural world. Particularly if you are a pastor, worship leader, Sunday school teacher, or parent, we hope you will share these ideas with others as well. Here are some practical ideas for weaving science into the life of your family, classroom, or congregation.

Continue learning about the natural world. You can do this by visiting a science museum, walking through a nature preserve, or reading a book by a scientist explaining her work in lay language. Ponder what this encounter with nature tells you about God the Creator. Does it give you insight into God's character? Do you see connections to biblical themes or passages? When one pastor read about the amazing neural networks in the brain, he incorporated this information into a sermon on John 15 about the vine and the branches. Be sure to share what you have learned about God with others!

Worship services can be enriched by nature and science in many ways:

▶ Use liturgies, Scripture readings, and prayers related to nature. One person led his congregation in prayer by thanking God for everything in the sanctuary, from the beauty of the stained glass windows to the healthy spleens in the bodies of the congregation (the "spleen prayer" was remembered for a long time!).

▶ Sing about the natural world, incorporating both well-loved classic hymns like "How Great Thou Art" and contemporary Christian songs that refer to modern science, such as "God of Wonders Beyond Our Galaxy."

▶ Make banners for the sanctuary inspired by the beauty of the natural world, or use photos of nature on screen during worship. One church invited the congregation to send in their own photos of creation at the end of the summer and used them in a themed worship service with readings from the Psalms. Besides beautiful scenery, look for images from modern science, like the detail within a cell or a nebula seen through a telescope.

▶ Sermons can draw on science themes in the same way they draw on modern films, books, and current events. Adopt an attitude from the pulpit that science can be positive, interesting, and faith-enhancing. Young people will appreciate the connections to their lives and will be less afraid that science will challenge their faith.

Some of the best ways to engage science are found outside of worship services. See chapters 1 and 13 for suggestions for Sunday school classes and small groups. Keep science in mind when planning youth group activities or other church events. For instance, after a winter evening service have a local astronomer set up a telescope for star gazing and read aloud Psalm 19. Or plan the church picnic near a nature center or in a large park, so the congregation can go on a nature walk and end with a song of praise.

ADDITIONAL RESOURCES

Achtemeier, Elizabeth. *Nature, God, and Pulpit.* Eerdmans, 1992.

Gordon, Charles. *In Plain Sight: Seeing God's Signature throughout Creation.* Designed on Purpose, 2009. Also at www.designedonpurpose.com. Forty devotional readings with beautiful nature images.

Huyser-Honig, Joan. "Science and Faith in Harmony: Positive ways to include science in worship." Article at wor.li/1100 from the *Calvin Institute of Christian Worship,* 2009. Includes links to many worship resources.

The Ministry Theorem, at ministrytheorem.calvinseminary.edu. Includes many resources for pastors and ministry leaders, including essays by scientists on the theme "What I Wish My Pastor Knew."

Season of Creation, at www.seasonofcreation.com. Offers liturgies, visuals, sermon themes, children's messages, and other worship resources on creation.

Wiseman, Jennifer. "Science as an Instrument of Worship: Can recent scientific discovery inform and inspire worship and service?" BioLogos white paper, 2009. www.biologos.org/uploads/projects/wiseman_white_paper.pdf.

APPENDIX

A SPECTRUM OF VIEWS ON ORIGINS

Likely you will encounter many views on origins. To give you an idea of the wide spectrum of viewpoints, we'll describe a number of theistic views. In these views God always plays a role.

▶ *Ancient Flat Earth.* A fully literal reading of Genesis 1-2 and other Old Testament passages describes a flat earth with a solid-dome firmament above the sky holding back the "waters above the earth." This is how the Old Testament era Hebrews and surrounding cultures pictured the world.

▶ *"Modern" Flat Earth.* The earth is flat but without the firmament or waters above. Genesis 1-2 and other Scripture passages are interpreted to require belief in a flat earth fixed in place; but words referring to the solid firmament and waters above the earth are interpreted differently.

▶ *Geocentrism.* The earth is spherical but fixed in place. Genesis 1-2 and other Scripture passages (Ps. 93:1; Josh. 10:12-13) are interpreted to mean that the earth doesn't move. The sun, moon, planets, and stars all move around the Earth.

▶ *Young-Earth Creation.* The modern sun-centered picture of the solar system is accepted as true, but the scientific picture of geological and biological history is disputed. Genesis 1-2 is interpreted as recent literal history; the earth and the universe are a few tens of thousands of years old. References to the firmament and waters above the earth are interpreted in a variety of ways. Although some "appearance of maturity" was included in creation—such as light from distant stars already on its way to earth—proper scientific measurements are thought to yield evidence that the earth and life on earth were recently created.

▶ *Young-Earth Creation: Created with Apparent Age.* Genesis 1-2 is interpreted as recent literal history; the earth and the universe are about ten thousand years old. But the universe and the earth were made to "appear" several billion years old, so scientific experiments measure only apparent age, not actual age.

▶ *Young-Earth Creation: Apparent Age Due to the Fall.* Genesis 1-2 is interpreted as recent literal history; the earth and the universe are about ten thousand years old. However, either because of the fall of man or the fall of Satan, the earth now appears much older.

▶ *Progressive Creation with Recent Creation of Earth and Life.* Genesis 1-2 is interpreted as recent literal history—but just for our planet and the creatures on it. The universe itself is billions of years old, following the evidence of astronomy.

▶ *Progressive Creation with Special Creation of New Life-forms.* The earth and the universe are several billion years old. At various times during biological history, God performed distinctive miracles to specially create each new life-form. Species have not descended from a common ancestor.

▶ *Progressive Creation with Common Ancestry and Modification.* The earth and the universe are several billion years old. All life-forms are linked by common ancestry, and some microevolution took place. However, at various times during biological history, God also performed distinctive miraculous acts in order to give certain life-forms new features or greater complexity. God might have altered existing species or perhaps worked through a sort of miraculous genetic engineering.

▶ *Progressive Creation Through "Miraculous" Evolution.* God used evolution, but the success of evolution is scientifically "surprising." Life-forms have changed and become much more complex than would be expected by the mechanisms of evolution alone. God must have been directing the evolutionary process, perhaps arranging for the process to travel along preordained paths, leading to much-better-than-expected outcomes.

▶ *Evolutionary Creation with Special Creation of First Life.* The history of life on earth happened as described by the theory of evolution, with nothing surprising about its success. God created, and evolution was the tool he used. However, the fact that biological evolution got started in the first place cannot be explained by science. The very first life on earth must have been miraculously created.

▶ *Evolutionary Creation.* The history of life on earth happened as described by the theory of evolution. God designed the natural laws of the universe to be just right for first life to assemble and for biological evolution to happen. These are natural processes that God governs, just like every other natural process. God's governance of these natural processes is pictured in a variety of ways:

  ▶ *Evolutionary Creation with Programmed Outcome.* The natural laws that govern evolution are designed to ensure that only certain kinds of life-forms will evolve. God ordained and intended our existence and designed natural processes to achieve more or less just what we see today.

  ▶ *Evolutionary Creation with Chosen Outcome.* Biological evolution could, in theory, have followed many different paths with different outcomes. However, the exact path that evolution took on earth and the final outcome we see today were entirely ordained by God, since every event that appears to be "random" to us is actually determined by God.

▶ *Evolutionary Creation with Flexible Outcome.* The exact path that evolution took on earth and the final outcome we see today were not entirely predetermined by God; rather, God gave his creation a certain degree of freedom. God also knew that this process would eventually produce intelligent, personal creatures to whom God could reveal himself.

▶ *Evolutionary Creation Known Only Via Special Revelation.* God designed and created the laws of nature so that life would evolve. We can't learn much about God's governance simply by studying the natural world. Nevertheless, we believe that creation occurred through God's hand because of God's special revelation in Scripture.

▶ *Deistic Evolution Plus Divine Involvement with Humans.* God created the universe and the laws of nature and then set them in motion without any intervention or meaningful governance. God got more involved with the world once humans came along.

INDEX

Developmental biology: 197
Dialogue on the Two Principal World Systems: 93
Differential reproductive success: 181, 184
Divine action/supernatural miracles: 15, 45, 49-50, 189-190
Divine suzerain: 134
DNA: 181, 184, 214
Doctrines of creation: 99
Dynamic stability: 46
Dynamic universe: 157

Ecological niche: 199
Ecosystem: 199
Eddy, John: 120-122
Egyptian cosmology: 137
Electrons: 174
Elementary particles: 174
Enuma Elish: 138-139
Evidence of age of universe: 161-165
Evidence of age of universe from asteroid orbits: 164
Evidence of age of universe from meteorites: 163
Evidence of age of universe from star clusters: 164
Evidence of expansion of universe: 166
Evidence of fusion at beginning of universe: 168
Evolution/Intelligent Design in the media: 186, 193, 211, 224
Evolution/theory of evolution: 11, 16, 178-180, 184, 186-188, 193
Evolution of complexity: 217
Evolutionary creationism: 16, 27, 180, 188-190, 206, 224, 236
Evolutionism: 16, 184-188, 193, 224, 231
Expansion rate of universe: 171-172
Experience altering and improving interpretation of Scripture:
    30-31
Experimental method: 15, 58
Experimental science: 59
Experimental variables: 59
Explainable natural events: 15

Whitcomb, John: 118
Woodward, John: 104-105
Worldview: 39
Worldviews and science as mutual influencers: 75
Worldviews and science—importance of evaluating statements on their own merit: 77
Worldviews held by scientists: 40-42
Worldviews/science in conflict: 76
Worldviews/scientists in cooperation: 15, 41, 76
Worship/praise/wonder in context of origins/scientific knowledge: 17, 33, 49, 289-291

Young-earth creationism: 27, 100, 102, 104-105, 114, 118-120, 123-124, 187, 224